VERY HIGH MULTIPLICITY PHYSICS WORKSHOPS

Proceedings 2000-2007

VERY HIGH MULTIPLICITY PHYSICS WORKSHOPS

Proceedings 2000-2007

edited by

Aleksey Sissakian

Joseph Manjavidze

Joint Institute for Nuclear Research, Russia

NEW JERSEY • LONDON • SINGAPORE • BEIJING • SHANGHAI • HONG KONG • TAIPEI • CHENNAI

Published by

World Scientific Publishing Co. Pte. Ltd.
5 Toh Tuck Link, Singapore 596224
USA office: 27 Warren Street, Suite 401-402, Hackensack, NJ 07601
UK office: 57 Shelton Street, Covent Garden, London WC2H 9HE

British Library Cataloguing-in-Publication Data
A catalogue record for this book is available from the British Library.

VERY HIGH MULTIPLICITY PHYSICS
Proceedings of the VHM Physics Workshops

ISBN-13 978-981-283-494-2
ISBN-10 981-283-494-X

Printed in Singapore.

PREFACE

The very high multiplicity (VHM) physics was devised to investigate cold and dense hadron medium similar to the neutron stars. The point is that in the VHM processes the kinetic energy of incident particles dissipates intensively and the final state of interacting media becomes cold and dense since the high virtuality partons don't leave the interaction area.

The VHM physics is a new stage of high energy hadron physics and the laboratory investigation of hadron media under such an extremal condition encounters a number of experimental and theoretical problems.

Experimental problems are as follows:

- Low value of the cross section (less than a few *pb*);
- Large number of produced hadrons (tens of thousands of charged particles in the heavy ion collisions);
- Low momenta of produced hadrons (comparable with pion mass).

Theoretical problems are as follows:

- The perturbative Leading Logarithm Approximation of QCD can not be applied since the momenta of partons of interacting media are comparatively small;
- The parameters of VHM processes exceed the range of validity of known theoretical models (Regge model, QCD string models, etc.).

The investigation of VHM processes could stimulate new experimental and theoretical methods. In this way it stood to reason to invite to the Workshops both theorists and experimentalists as well as experts in various fields of hadron physics.

Collective phenomena and coloured plasma (QGP) state are being searched intensively at various laboratories. Some definite evidence of existence of QGP has been produced in recent experiments at the relativistic ion collider RHIC in BNL (USA). The investigations of dense and hot hadron media allow to view the problem of extremal hadron media from another angle.

Planned experiments at NICA (Nuclotron-based Ion Collider Facility, JINR) was intended to research above-stated issues and here we start with the corresponding papers. We hope that the articles in the present book will give sufficient information concerning the problem of dense hadron media and will originate further investigations.

A. Sissakian and J. Manjavidze

CONTENTS

NUCLOTRON-BASED ION COLLIDER FACILITY (NICA)

I. MESHKOV, A. SISSAKIAN and A. SORIN

Joint Institute for Nuclear Research,
Joliot Curie st., 6, Dubna, Moscow region, 141980 Russia

The project of an ion collider accelerator complex NICA that is under development at JINR is presented. The article is based on the Conceptual Design Report (CDR)[1] of the NICA project delivered in January 2008. The article contains NICA facility scheme, the facility operation scenario, its elements parameters, the proposed methods of intense ion beam acceleration and achievement of the required luminosity of the collider. The symmetric mode of the collider operation is considered here and most attention is concentrated on the luminosity provision in *collisions of uranium ions (nuclei).*

1. Introduction

General goal of the project is to start in the coming 5-7 years experimental study of hot and dense strongly interacting QCD matter and search for possible signs of the mixed phase and critical endpoint in heavy ion collisions[2] . The Nuclotron-based Ion Collider fAcility (NICA) and the MultiPurpose Detector (MPD)[3] are proposed for these purposes.

Another goal of the NICA construction is performance of experimental studies on spin physics with colliding beams of polarized protons and light nuclei. For this purpose the second collision point is foreseen and a dedicated detector is planned to be constructed as well.

One of key elements of NICA is superconducting (SC) proton synchrotron Nuclotron constructed and being operated in JINR. The design of the new accelerator complex NICA is based on the experience obtained during the Nuclotron development and is using modern concepts of accelerator technique. As a result of the project realization, the potential of the Nuclotron accelerator complex will be sufficiently increased in all the fields of its current physics program and the facility creation will expand these experimental studies in much higher energy range. That makes NICA a unique research tool.

The choice of an optimum facility structure and parameters was based on the following criteria:

-the range of the center-of-mass energy available for experiments is of $\sqrt{s_{NN}} = 4 - 11$ AGeV,

-the sufficient luminosity level for collisions of heavy ions has to be provided up to Uranium,

-the choice of technical solutions for NICA design and construction has to be made avoiding as much as possible a long R&D stage,

-the technologies developed at JINR and experience of the JINR personnel has to be applied as much as possible,

-an optimal use of the existing equipment and the existing buildings for the facility dislocation is assumed,

-wide co-operation with institutions from JINR Member States (and Russian institutions especially) is foreseen,

-the project cost has to be optimized in accordance with the JINR budget increase in coming years.

The project development includes the following tasks:

-Upgrade of the Nuclotron aiming to reach its parameters required for NICA operation,

-Development of highly charged heavy ion source and construction of a new one based on the principle of The Electron String Ion Source,

-Design & construction of a new linear accelerator as an injector,

-Design & construction of a new booster synchrotron equipped with electron cooling system,

-Design & construction of two new superconducting storage rings equipped with electron and/or stochastic cooling systems to provide *the collider experiment* with heavy ions like Au, Pb and U in the energy ion of 1 - 4.5 GeV/u $\left(\sqrt{s_{NN}} = 1 - 11\ GeV/u\right)$ at the average luminosity of 10^{27} cm^{-2}·s^{-1} (R&D stage for the cooling systems is required). At *the second stage of the project development* the colliding beams of light ions and polarized protons and deuterons in the total energy range available with the Nuclotron will be provided (R&D stage for polarized particles acceleration is required). This allows one to continue at JINR the traditional program of experimental studies raising it to a new level. The NICA project assumes also a collision of mass asymmetric beams including proton-ion collisions. It is quite important as complimentary physics experiments giving a reference point for comparison with heavy ion data. In future, as a next stage of the project, the NICA keeps a possibility for electron-ion collisions.

The maximum magnetic rigidity of the collider rings is chosen to be equal to the Nuclotron one, i.e. 45 T·m. The intensity of polarized beams is planned to be at the level of $1{\cdot}10^{11}$ particles per beam.

An essential feature of the project is the use of two boosters – the Booster-synchrotron and the Nuclotron. This booster tandem is intended to accelerate ions and to fill collider rings which operate at chosen experiment energy from the energy range indicated above. Such a scheme gives essential advantages for the experiment performance in the colliding beams mode.

General challenge of heavy ion collider experiment is to achieve a high luminosity level in a wide energy range beginning with about 1 GeV/u. To reach this goal the beam intensity has to be of about $2{\cdot}10^{10}$ ions in each collider ring distributed over 10–20 bunches.

The collider rings have two interaction points (IP). In one of them the Multi Purpose Detector (MPD) [3] will be located. The design of this IP provides collisions at *zero crossing angle*. The second IP is reserved for another detector that will be designed and constructed at the next stage of the project development. One of its versions is the Spin Physics Detector (SPD) dedicated to experiments with polarized protons and deuterons at the maximum value of $\sqrt{s} = 27.2\ GeV/u$ for protons and 13.7 GeV/u for deuterons.

2. The NICA structure and operation scenario

The proposed facility (Fig. 1, 2 and Table 1) consists of the following elements:

1. *ESIS-type ion source.* It provides U^{32+} ions at intensity of $2{\cdot}10^9$ ions per pulse of about 7 μs duration at repetition rate up to 50 Hz.

2. *Linear accelerator*: RFQ and RFQ Drift Tube Linac (RFQ DTL). The linac accelerates the ions of mass to charge number ratio $A/q \leq 8$ up to the energy of 6 MeV/u at efficiency no less than 80%.

3. *Booster synchrotron* of the maximum magnetic rigidity of 25 T·m and of the circumference of 215 m. It is equipped with electron cooling system that provides cooling of the ion beam in the energy range from injection energy up to 100 MeV/u; the maximum energy of U^{32+} ions accelerated in the Booster is of 440 MeV/u.

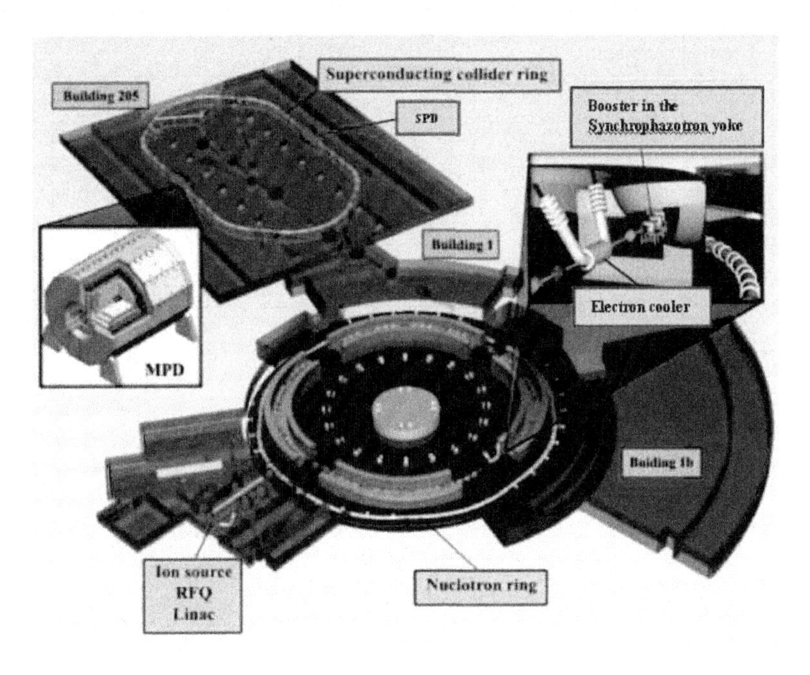

Fig. 1. NICA location in the existing buildings

4. *Stripping target station* placed in the transfer line from the Booster to the

Nuclotron. Its stripping efficiency at the maximum Booster energy is no less than 40%.

5. *The Nuclotron*: SC proton synchrotron of maximum magnetic rigidity of 45 T·m and of the circumference of 251.52 m. It provides the ion acceleration up to the experiment energy range of 1-4.5 GeV/u for U^{92+}.

6. *Two collider rings* with maximum magnetic rigidity of 45 T·m and the circumference of about 225 m. The maximum field of superconducting dipole magnets is of 4.5 T. For luminosity preservation during an experiment performance an electron and or stochastic cooling systems is to be constructed.

The collider ring circumference is chosen to be compatible with the existing building #205 (Fig. 1).

Table 1. General parameters of the NICA rings ($U \times U$ collisions)

	Booster (U^{32+})	Nuclotron (U^{92+})	Collider ($U^{92+} \times U^{92+}$)
Ring circumference, m	216	251.52	224
Injection energy, MeV/u	6.2	440	1000-4390
Final kinetic energy, GeV/u	0.44	1.0-4.39	1.0-4.39
Magnetic rigidity, T ·m	2.4-25	8.2-45	14.5-45
Bending radius, m	14	22	10
Dipole magnetic field, T	0.17–1.5	0.37–2.05	1.6–4.5
Number of dipole magnets	40	96	24
Number of quadrupoles	48	64	32
Magnetic field ramp, T/s	2.65	1.27	> 25
Pulse repetition rate, Hz	1	1	0
RF harmonics number	4/1	1	90
RF frequency range, MHz	0.6–1	0.857–1.17	105–117
RF voltage, kV	4	15	100
Residual gas pressure (Nitrogen equivalent at room temp.), Torr	10^{-11}	10^{-8}	10^{-10}

The facility *operation scenario* (Fig. 2, 3) is aimed to provide generation of a new ion bunch and its injection into one of the collider rings during every 4–5 seconds to fill both collider rings for less than 3 minutes. The collider rings are filled ones at the beginning of experiment and the beams (30 bunches) have to be refreshed after a long time about 1 h that depends on the intrabeam scattering process and the beam cooling system parameters (section 5.5).

The magnetic field ramp duration of the Booster includes about 1 s for the ion storage at injection and acceleration to 50 (100) MeV/u, 1 s of the electron cooling, 1 s for the acceleration to 400 MeV/u. The RF voltage amplitude corresponds to the bunch acceleration. The bunch compression system in Nuclotron will be operated at about 120 kV.

The number of injections into the Booster does not influence practically on the bunch preparation time because of the high repetition frequency of the ion source. Therefore the ion bunch of intensity of a few times larger (if necessary) can be prepared during practically the same time.

The bunch preparation in the Nuclotron includes the bunch acceleration and the bunch compression at the plateau of the magnetic field. The total bunch preparation time is defined mostly by the duration of acceleration and electron cooling in the Booster, therefore it is almost independent of the ion specie and the experiment energy.

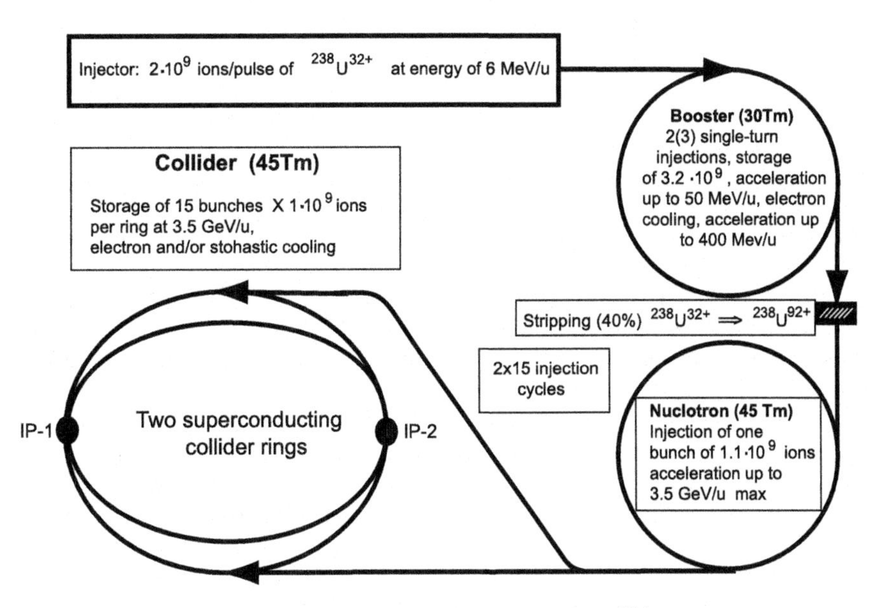

Fig. 2. Scheme of the NICA with parameters for collisions of U^{92+} ions at the energy of 3.5 GeV/u

3. Luminosity of heavy ion collisions

Main effects, which define the collider luminosity value, have been taken into account at the collider design.

1. Beta function value in the interaction point chosen to be equal to 0.5 m and the developed rings lattice provides it when the maximum beam radius in the lenses of the low beta insertion section is about 4 cm that requires reasonable

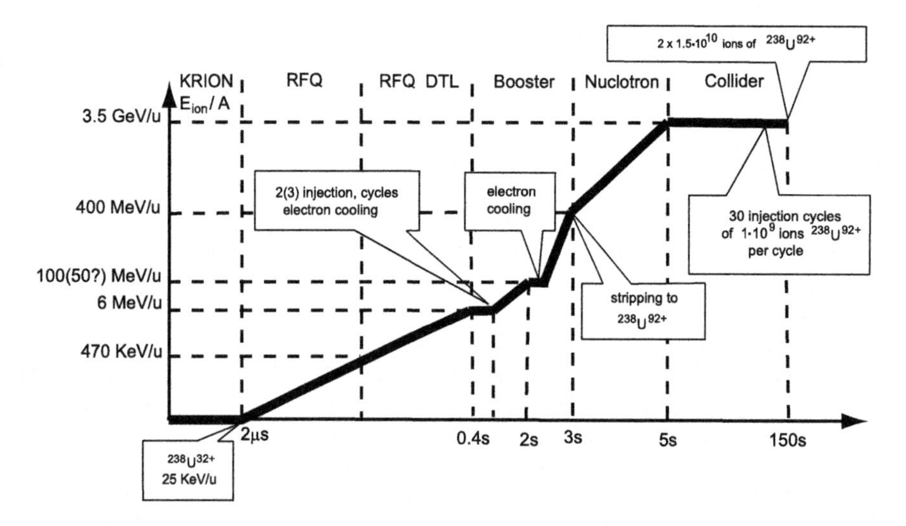

Fig. 3. Time table of the ion storage process

aperture of the lenses.

2. Short bunch length of about 30 cm rms makes possible to avoid "the hour glass effect" and to concentrate 80% of the luminosity inside the inner tracker of the MPD.

3. Collider will have to be operated at the beam emittance corresponding to the space charge limit. In the NICA energy range the luminosity is limited by the incoherent tune shift value. If the ion number per bunch and the tune shift are fixed the luminosity is scaled with the energy as $\beta^3\gamma^3$. The formation and preservation of low emittance value, corresponding to achievable tune shift, is produced by beam cooling application at the experiment energy (section 5.5).

4. Large collision repetition rate is provided with the collider operation at the bunch number of 10÷20 in each ring. This is achieved by the injection kicker pulse duration of about 100 ns and by injection into the collider of bunches of a sufficiently short length. The bunch of the required length is formed in the Nuclotron after the acceleration. Small longitudinal emittance value, required for the bunch compression in the Nuclotron, is provided by the electron cooling of the ion beam in the Booster.

5. Sufficiently long luminosity life-time will be achieved with an application of the electron and/or stochastic cooling systems. In equilibrium between intrabeam scattering (IBS) and the cooling the luminosity life-time is limited mainly by the ion interaction with the residual gas atoms and recombination of the ions with the electrons in the electron cooler. The vacuum conditions in the collider rings are chosen to provide the beam life time of a few hours. The beam preparation time is designed to be between 2 and 3 minutes (Fig. 3). Therefore, the mean luminosity value should be close to the peak one.

Finally the beam and rings parameters allow reaching the required luminosity (Table 2).

An important question is *the luminosity dependence of the colliding ions energy.* As is seen from the data of Table 2, the dominated *space charge effect in colliding beams* is related to incoherent tune shift ΔQ, where as beam-beam effect (parameter ξ) is practically negligible. Therefore, when collision energy is tuned to the low values, one should take care of incoherent tune shift magnitude keeping it below certain value:

$$\Delta Q \quad \leq 0.05 \ .$$

Table 2. NICA parameters for U-U collisions

Parameter	Value
Number of interaction points	2
Beta function in the collision point, m	0.5
Rms momentum spread	0.001
Rms bunch length, m	0.3
Number of ions in the bunch	$1 \cdot 10^9$
Number of bunches	15
Incoherent tune shift, ΔQ	0.05
Beam-beam parameter, ξ	0.002
Rms beam emittance (unnormalized), $\pi \cdot$ mm $\cdot$ mrad at 1 GeV/u at 3.5 GeV/u	 3.8 0.26
Luminosity per one interaction point, $cm^{-2} \cdot s^{-1}$ at 1 GeV/u at 3.5 GeV/u	 $1.7 \cdot 10^{26} (L_2)$ $1.1 \cdot 10^{27}$

Analyzing the Formulae for ΔQ, ξ and luminosity L one can draw the definite conclusion about the luminosity variation versus energy. Indeed, for the Gaussian cylindrical colliding beams we have

$$\Delta Q = \frac{Z^2}{A} \cdot \frac{r_p N_{bunch}}{4\pi\beta\gamma^2\varepsilon_{norm}} \cdot k_{bunch} \ , \tag{2}$$

$$\xi = \frac{Z^2 r_p}{A} \frac{N_b}{4\pi\beta\varepsilon_{norm}} \frac{1+\beta^2}{2} \ , \tag{3}$$

$$L = \frac{\beta^2 \gamma N_{bunch}^2}{4\pi\varepsilon_{norm}\beta^*} \cdot \frac{C_{Ring}}{c} n_{bunch} \cdot F\left(\frac{\sigma_s}{\beta^*}\right) \ .$$

Here Z, A are atomic number and atomic mass number of colliding ions, r_p is proton classic radius, N_{bunch} is ion number in a bunch, ε_{norm} is normalized emittance of the bunch,

$$k_{bunch} = \frac{C_{Ring}}{\sqrt{2\pi}\sigma_s} ,$$

k_{bunch} is the bunching factor, C_{Ring} is the collider circumference, σ_s is the rms bunch length (Gaussian constant), n_{bunch} is bunch number circulating in the collider ring and the function $F(x)$ is given by the following formula:

$$F(x) = \frac{1}{\sqrt{\pi}} \int_{-\infty}^{\infty} \exp(-u^2) \cdot \frac{du}{1+(ux)^2}, \quad F(x) \sim 1 \;\; when \;\; x << 1 .$$

Assuming a small variation of the normalized emittance ε_{norm} with ion energy in the collider one can see (Formula (1)) a week dependence of beam-beam parameter ξ of energy. Then one can point out two outmost cases for luminosity dependence of energy *at constant* ΔQ.

1) The normalized emittance is const: ε_{norm} = const. In this case ion number per bunch N_{bunch} has to be scaled with energy (see Formula (1)) as

$$N_{bunch} \propto \beta\gamma^2 .$$

And luminosity depends on energy as

$$L_1 \propto \beta^4\gamma^5(\Delta Q)^2 . \tag{8}$$

2) Ion number per bunch N_{bunch} is defined by bunch preparation scenario (see sections 1 and 3) and is as large as possible, i.e. N_{bunch} and ΔQ are constant when energy is tuned. Then the bunch normalized emittance ε_{norm} has to be changed "artificially" when energy is tuned:

$$\varepsilon_{norm} \propto \frac{N_{bunch}}{\beta\gamma^2\Delta Q} ,$$

and the luminosity varies with energy as

$$L_2 \propto \beta^3\gamma^3 \cdot \Delta Q . \tag{10}$$

The plots of L_1 and L_2 versus energy (Fig. 4) show a week difference of two cases. So, if $L_1 = L_2$ at 3.5 GeV/u they decrease to $L_1 = 0.9 \cdot 10^{26}$ and $L_2 = 1.7 \cdot 10^{26} cm^{-2} \cdot s^{-1}$ at 1 GeV/u .

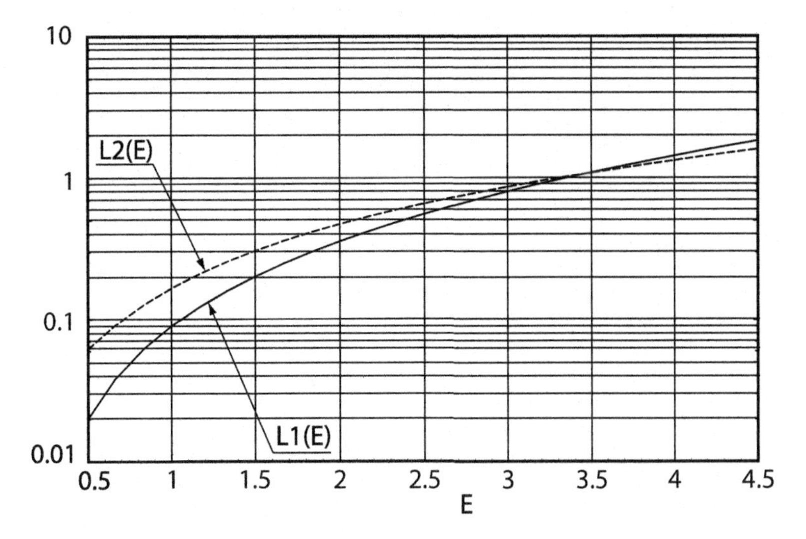

Fig. 4. Collider luminosity versus ion energy in two outmost cases (3) and (4). L in the units of $1 \cdot 10^{27} cm^{-2} \cdot s^{-1}$, E in GeV/u

4. Injection chain structure and efficiency

The structure of the NICA injection chain is similar to that one proposed for RHIC on the basis of EBIS-type ion source (Fig. 5). The NICA elements described below (section 5) provide the required ion bunch parameters injected into collider. Significant advantage of the proposed injection chain is the electron cooling application of the ion beam at intermediate energy in the Booster. Total efficiency of the injection chain from the ion source to the collider is assumed to be in the range from 20 to 25%.

5. The main elements of NICA

5.1. *Injector*

The existing Nuclotron injection complex consists of HV fore-injector at 700 kV pulsed potential and Alvarez-type linac LU-20. The injector delivers protons at the energy of 20 MeV and ions at $Z/A \geq 0.33$ at the energy of 5 MeV/u[4] . The wide range of the ion species is provided by the heavy ion source "KRION-2", duoplasmatron ion source, polarized deuteron source POLARIS and laser ion source.

The new injector will be constructed under NICA project. It consists of the following elements:

-Electron String Ion Source (ESIS) that delivers the heavy ions at the charge state of 30+÷32+ and the heavy ions like Au^{51+} and U^{64+} at the intensity of $(2{\div}4){\cdot}10^9$ ions per pulse of 7 μs duration (the ion revolution period in the Booster),

-External uranium ion source,

-Source of polarized D^- ions (POLARIS or CIPIOS),

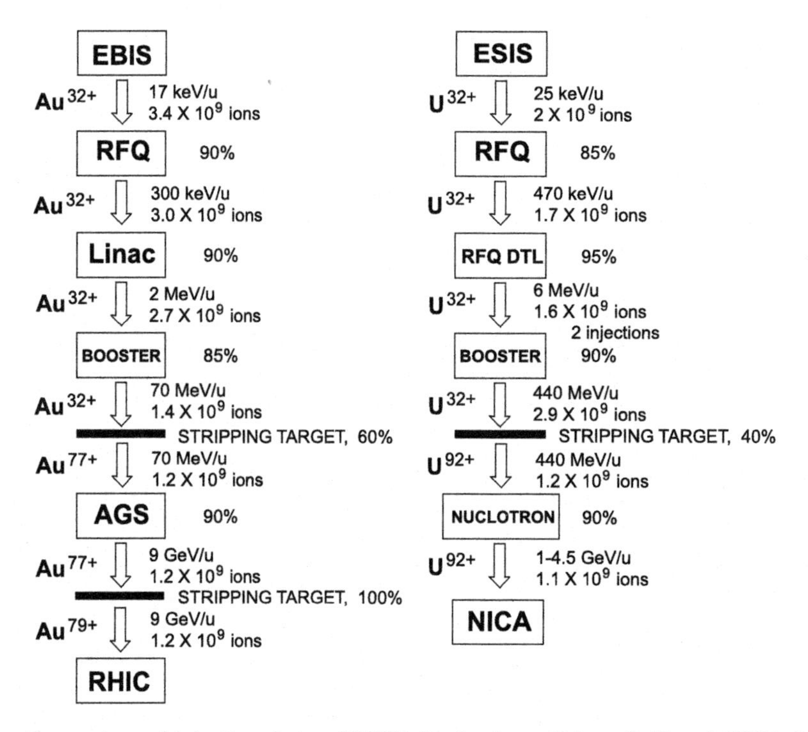

Fig. 5. Comparison of injection chain of RHIC for Au-Au collisions (left) and NICA for U-U collisions (right)

-Low Energy Beam Transport (LEBT) from the source to the entrance of the linear accelerator

-Linear accelerator

-High Energy Beam Transport (HEBT) from the exit of the linac to the Booster; this line contains the debuncher.

A reserve option of the injector complex anticipates operation with an Electron Cyclotron Resonance Source (ECRS) of heavy ions that assumes the multiturn injection into the Booster.

For generation both heavy ions and polarized light ones the injector linac has to fit to the following requirements:

-acceleration of the heavy ion beam with the parameters listed for ESIS above,

-acceleration of the heavy ions at the charge state of about $30+\div 32+$ at the current of $0.1\div 0.2$ mA and pulse duration up to 100 μs (operation with ECRS),

-acceleration of polarized D^- ions at the current up to 1 mA and the pulse duration of about 1 ms (operation with POLARIS or CIPIOS).

This report, as mentioned above, is dedicated to description U-U collision experiment. Therefore, the first option (ESIS) is described in here.

Electron string Ion Sourse (ESIS) is an advanced version of the Electron Beam Ion Source (EBIS) that has been invented and developed in pioneering works at

JINR[5] and is used widely in many accelerator centers for production of highly charged ions or bare nuclei. Electron beam ion sources are operated in a pulsed mode and they are used mainly at injectors of ion synchrotrons. It is well known that the EBIS-type ion sources produce the highest pulsed current of highly charged ions, but the average ion beam intensity is rather low. One of the methods of increasing the ion yield from EBIS is to increase the electron beam current in the source, as well as the current density. To avoid space charge problems in the EBIS electron beam one needs to increase the electron beam energy. This leads to decrease of ionization cross-section and significant enhancement of the beam DC power (sometimes up to hundreds kW). An application of so called "energy recovering mode" of electron beam operation in the EBIS facilitates the problem, but not radically.

A possible solution is so-called "reflex mode of EBIS operation" (Fig. 5) proposed at JINR in 1994[6] . It promised to decrease by hundred times a power of the electron beam at the same ion yield.

In the reflex mode of the EBIS operation the electrons do not reach electron collector after one pass through the drift space of the source. Instead they are reflected back towards the electron gun and are bouncing in such a way several times between the cathode and the reflector generating highly charged ions much more effective than a direct electron beam does. The gun cathode and reflector are placed in a fringe magnetic field (1/20 of a maximal field).

It was found later that in certain conditions the "cloud" of the repeatedly reflected electrons when confined in a strong solenoid magnetic field undergoes some kind of phase transition. This leads to a stepwise increase of the confined electron plasma density in a new steady state called by the authors "the electron string". Moreover, both the total number of electrons in the drift tube, and the total number of the produced ions grow up stepwise also. The transition usually goes via an unstable pre-string state in which the electron energy spectrum expands that further suppresses the instability[7] .

The fulfilled studies have shown that the EBIS version named as Electron String Ion Sources (ESIS) is a very efficient way of highly charged ions generation. Its very important peculiarity is a strong dependence of ion yield on the solenoid magnetic field B: the yield measured in the range of 0.5÷3.5 T nicely fitted to cubic law

$$n_i = a_i B^3 \ , \ n_e = a_e B^3 \ ,$$

where n_i and n_e are the numbers of delivered ions and the electrons confined in the electron string, a_e, a_i are the constants determined by source geometry parameters. Presently the new superconducting solenoid of B up to 6 T is being constructed and its application has to demonstrate in nearest future a capability of ESIS to meet the NICA project requirements.

The electron string studies carried out during recent years in a collaboration of JINR group with Manne Sighbann Laboratory (Stockholm, Sweden), Institute fuer Angewandte Physik, J.W. Goethe-Universitat (Frankfurt/M, Germany),

Brookhaven National Laboratory (Upton, USA) and Research Center for Nuclear Physics (Osaka, Japan) did confirm[8] that the electron string phenomenon is a new, unknown before, state of the strongly magnetized electron beam ("cloud") that can be used successfully for the highly charged ions generation.

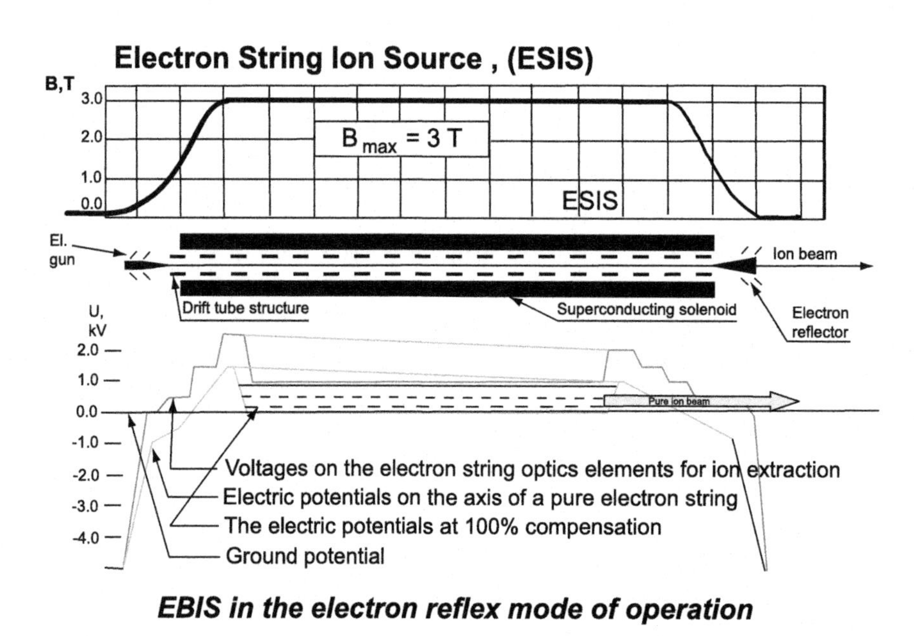

Fig. 6. Scheme of the Electron Beam Ion Source in the electron reflex mode of operation

Another peculiar feature of EBIS (and ESIS as well) is a very small ion beam emittance. It is caused by very small cross-section of the electron beam, which confides the ions. As result, the ESIS is capable to deliver an ion beam acceptable by the linac. Unfortunately, the ESIS beam emittance never been measured – this task is formulated as an urgent one in the NICA project plan. There is only an indirect result presently: at two Nuclotron runs during recent years the ESIS "Krion-2" (first version of ESIS) was used for delivery of the ions C^{6+}, N^{7+}, O^{8+}, Ar^{17+}, Fe^{24+}, Au^{51+} and all the beams were successfully accelerated in the linac LU-20 that has aperture of 15 mm radius. Only available data of direct emittance measurement have been obtained with the source of EBIS type at Brookhaven National Lab. The measured *normalized rms emittance* was of 0.07 π·mm·mrad.

External uranium ion source will be designed and constructed during the project development. This work was not started yet and is postponed presently.

The Source of polarized D^- ions POLARIS is being used at Nuclotron and will be replaced by more advanced version CIPIOS in the next year.

The Low Energy Beam Transport (LEBT) transfers the beam from the ESIS (placed on the 200 kV platform) to the linac. It is of 2.5 meter long and consists of an extraction and pre-acceleration system, spectrometer bending magnets and a focusing system for the transverse emittance matching. Two solenoid coils are used for this goal. The spectrometer magnets are used for injection of polarized D^- ions into the LEBT. Two or three quadrupole lenses are required for transformation of the "flat" D^- beam after the spectrometer in an axially symmetric one.

Linear accelerator will be designed and constructed by the collaborating group from IHEP (Protvino). Preliminary design based on the group long-standing experience[9] has been developed at the NICA CDR preparation. It was proposed to construct a linac consisting of front-end RFQ section and four sections with a spatially periodic RFQ focusing named as Drift Tube Linac (RFQ DTL) sections. Such a linac has definite advantages:

1. High capture efficiency and low injection energy,
2. Reduced dimensions,
3. Large peak current,
4. Technological simplicity, an ease of fabrication,
5. Small power losses related to wall current at low ion velocities.

Both *the RFQ* and *the RFQ DTL* sections are designed (Table 3) to work at the frequency of 75 MHz. The linac acceptance is of 0.5 π·mm·mrad and the beam emittance during acceleration grows not more than by two times. The matching of the axial symmetric DC beam from the ion source with the RFQ acceptance is done by adiabatic increase of the focusing gradient at the entrance of the RFQ. Large capture efficiency is provided by a choice of the synchronous phase at the entrance of the RFQ close to -90 $^\circ$. To avoid an increase of the accelerator length the synchronous phase value is decreased fast to certain value in the initial part of the RFQ (in so called "chopper" section). Main part of the RFQ accelerator works as an adiabatic buncher and the synchronous phase at the exit of the RFQ is equal to $-30\,^\circ$. This is equal to the synchronous phase in the RFQ DTL. The longitudinal matching of the bunch with the RFQ DTL is provided by the required choice of the acceleration rate at the RFQ DTL entrance.

Rather low electric field at the electrode surface (Table 3) allows to avoid a significant electron load of the cavity (cold emission).

Sufficient part of the accelerator cost is expenses of the RF generators. All the amplification cascades feeding the linac sections will work with the same setting generator. The electronic tube GI-27AM of Russian production will be used in the powerful intermediate and final cascades. Maximum power and anode voltage of each tube are of 5 MW and of 32 kV respectively. The anode voltage modulators will be constructed with semiconductor electronics elements.

High Energy Beam Transport (HEBT) from the linac to the Booster contains the debuncher that consists of two single-gap cavities of the 148.5 MHz frequency. One of them is located at the exit of the linac, another one near the Booster entrance. The HEBT contains two or three bending magnets and about 10 quadrupole lenses, which belong to the Amsterdam Pulsed Stretcher transferred from NIKHEF, Amsterdam, to JINR in 1999[10] . Magnetic rigidity of these dipole magnets is about

Table 3. General design parameters of the linac

Parameter	RFQ	RFQ DTL
Number of sections	1	4
Input energy, keV/u	200 / 8	473
Output energy, keV/u	473	6200
Max. peak surface E-field, kV/cm	317	350
Aperture radius of channel, mm	3.01	4.5 ÷ 5.5
Transverse acceptance, $\pi \cdot mm \cdot mrad$	70 unnormalized	2.26 normalized
Overall length, m	4.05	13.6
Accelerating efficiency, %	86	

4.2 T·m and the bending radius of 3.3 m that is sufficient for the beam steering.

5.2. *Booster synchrotron*

The main functions of the intermediate heavy ion synchrotron, the Booster (Table 1), are the accumulation of 4•10^9 U^{32+} ions and acceleration them up to the energy of 440 MeV/u that is sufficient for stripping of these ions from 32+ charge state to 92+ one. The last operation simplifies significantly the requirements to the vacuum conditions in the Nuclotron due to essential decrease of ions recombination rate on residual gas atoms. The application of the electron cooling at the ion energy of 50÷100 MeV/u decreases the ion beam longitudinal emittance.

The present layout of the Nuclotron and existing injection and extraction lines make possible placing the 216 m Booster inside the Synchrophasotron yoke (Fig.1).

Two versions of *the Booster lattice* are being considered - DFO and FODO periodicity. The first one permits to make more efficient collimation of charge exchanged ions whereas the second is more convenient at synchrotron tuning and operating. At the same time the low beam intensity (below 4•10^9 ions per pulse) allows us to install a small number of collimators to clean out the U^{33+} ions from the U^{32+} beam. That is why the FODO lattice is considered as more preferable for further consideration.

The magnetic system of the Booster consists of 4 quadrants each one containing 10 dipole magnets, 6 focusing and 6 defocusing lenses, the multipole correctors for compensating the errors of both main (dipole, quadrupole) and higher (sextupole, octupole) harmonics of the magnetic field. The maximum magnetic field in dipoles is of 1.5 T (Table 1).

Two versions of the magnetic system - normal conducting or superconducting ones – have been considered. It was shown that both can meet the requirements for the Booster parameters. Finally the SC version has been chosen. One of its main advantages is a possibility to use so called "Nuclotron-type magnets", or so called "superferric" magnets[11] . Recently this type of magnets has been accepted as the basic one for SC magnetic system of the SIS-100 synchrotron in the FAIR project at GSI. It allows one combining international efforts on construction of such magnets. Technical details of the magnets design are described in the CDR [1].

Beam injection into the Booster is performed by the twice repeated single turn injection pulses of 7 μsec duration each one and of 0.1 sec interval between the pulses. The scheme of betatron stacking in the horizontal plane is applied with one septum magnet and three bump ones. *Fast (single turn) extraction* transfers the beam into the Nuclotron and, when necessary, to the experimental hall for a fixed target experiment performance. The extraction system consists of a kicker and septum magnets.

The Booster cycle is of 3 s period (Fig. 6) and has four parts:

1) adiabatic trapping at fixed frequency on magnetic field plateau,

2) ion acceleration at the 4th harmonics of the revolution frequency up to 50 (100) MeV/u and debunching,

3) electron cooling and beam bunching at the 1st harmonics of the revolution frequency,

4) ion acceleration at the 1st harmonics of the revolution frequency up to 440 MeV/u.

The RF system of the Booster works at two different regimes (Table 4).

Table 4. Main parameters of RF system

Parameter	1st stage	2nd stage
RF frequency, MHz	0.640/2.400	0.600/1.00
Harmonics number	4	1
Vmax/Vmin (kV)	7/1	7/1

The nearest prototype of the Booster acceleration system is that one designed by Budker INP (Novosibirsk) for SIS-100. It can be constructed on the base of GU92-A electronic tube with the cavities of ferrite loaded type.

The space charge effects in the Booster ion beam are related mainly with the incoherent tune shift (1). As the analysis shows the storage of $4{\cdot}10^9$ of U^{32+} ions at the energy of 6 MeV/u gives $\Delta Q_{max} \leq 0.05$ (see details in [1]).

The Booster electron cooler is applied mostly to decrease the longitudinal emittance from the injection value of about 7.5 eV·s to the necessary value of 2.5 eV·s. Cooling time is limited by the Booster operation cycle: it can not exceed the *value of 1 sec.* For the transverse plane the cooling system has to keep the magnitude of the normalized transverse emittances at the level of 1 π·mm·mrad (rms). It can be done by a proper misalignment of electron and ion beam axes on the level of about 1 mrad in both transverse planes (Fig. 7).

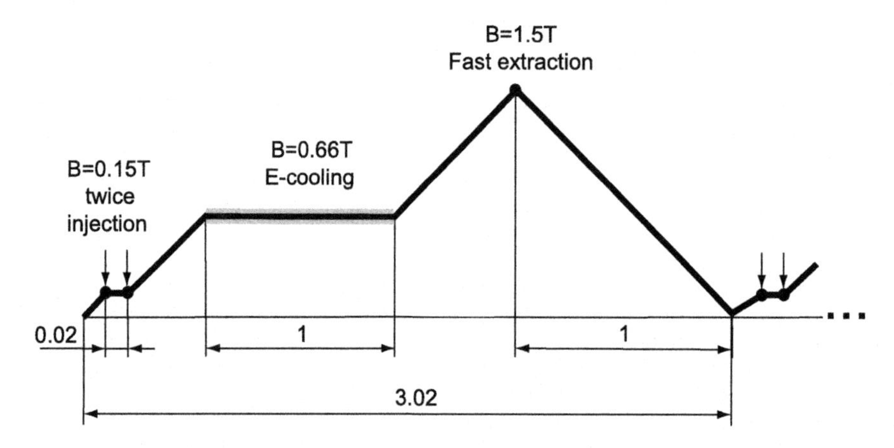

Fig. 7. The Booster cycle diagram

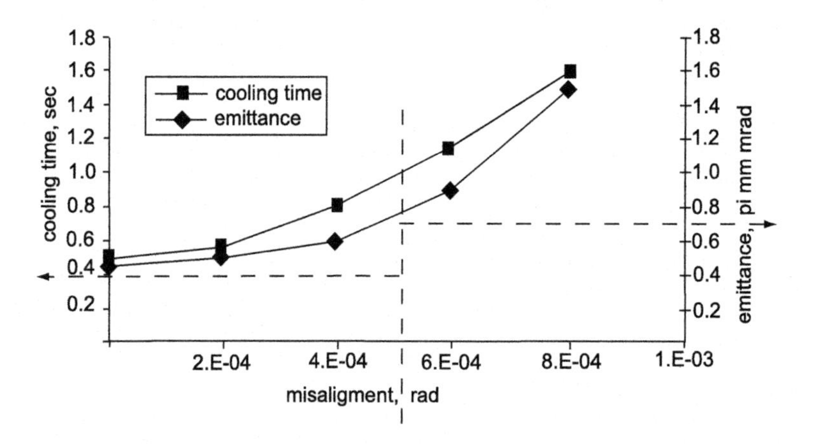

Fig. 8. The dependence of the cooling time and transverse emittance of the cooled ion beam on the misalignment angle between electron and ion beams axes; "the cooling time" is defined as a time interval when the longitudinal emittance decreases from 7.5 eV·s to 2.5 eV·s.

The optimal value of electron cooling energy has been chosen accounting many effects: beam lifetime limited by interaction with the rest gas, and recombination on the cooling electrons, space charge effects in the cooled ion beam, cooling time duration ($\leq$1 sec), electron beam space charge effect on ion cooling, an optimal regime of the RF station and, at last, the cost of the electron cooler. A cooler of electron energy of 50 keV that fits to the Booster parameters can be constructed by a conventional ("standard") scheme[12] . The electron current of 1 A does meet the Booster specifications.

5.3. *Stripping target station*

Stripping target station is placed in the transfer line between the Booster and the Nuclotron. Its stripping efficiency at the maximum Booster energy is no less than 40%. As its prototype of the stripping target design used at RHIC has been chosen. The carbon foil used in there has the thickness of 120 μm that is somewhat larger than the optimum one. Such a value was chosen to provide a long life-time of the foil.

An increase of transverse beam emittance at crossing the foil can be negligible if the beta function in the foil position is of a few meters. The longitudinal emittance grows mainly due to large foil non-uniformity. This effect can be suppressed by application of electron cooling in the Booster where the bunch length is sufficiently shortened.

5.4. *The Nuclotron*

The SC proton synchrotron Nuclotron (Fig. 8) has been built during 1987−1992[13] and is the world's first synchrotron based on low-field electromagnets with iron yoke and superconducting coils (see section 4.2). It accelerates ions up to the experiment energy in the range of 1÷4.39 GeV/u (for U^{92+}, Table 1). Main part of the Nuclotron systems has been fabricated by the machinery workshops of JINR and VBLHE without involvement of a specialized industry.

Fig. 9. The Nuclotron in the tunnel

The energy of accelerated particles achieved at the Nuclotron presently is 5.7 GeV for protons and 2.2 GeV/u for heavy ions (Fe, A = 56). The goal of nearest development of the Nuclotron is reaching of design parameters and, particularly, of 45 T·m magnetic rigidity (Table 1). The Nuclotron upgrade program (the "Nuclotron-M project") was started in 2007.

An essential upgrade will undergo the vacuum system of the Nuclotron that will have to lead to a sufficient improvement of the vacuum conditions in the Nuclotron ring.

Another urgent task is refurbishment of its RF system. It has to be capable acceleration of single bunch beam and its compression before extraction (Fig. 9). Nevertheless, the proposed RF system has reasonable parameters (Table 5).

Table 5. Main parameters of the RF system of Nuclotron for heavy ion (U^{+92}) acceleration and bunch compression

Number of particles	$1.1 \cdot 10^9$
Energy range, MeV/u	400/4390
RF frequency, MHz	0.98/1.3
Bunch rms longitudinal emittance (eV · s)	¡3
Harmonics number	1
RF amplitude at trapping (kV)	15
RF amplitude at lineal ramp (kV)	6.5
RF amplitude at flat top before compression (kV)	0.2
RF amplitude at compression (kV)	120
Magnetic field ramp, T/s	1
Max acceleration time, s	1.6
Compression time, ms	0.9

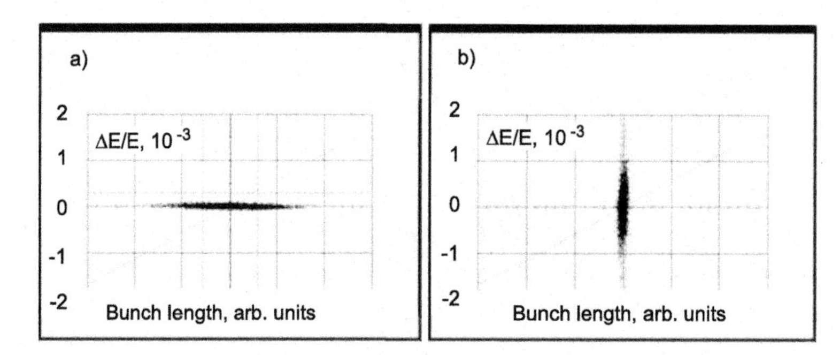

Fig. 10. Rotation of the U^{92+} ion bunch in the phase space: before (a) and after (b) compression procedure

5.5. *Collider*

Conception of the NICA collider (Tables 1, 2) accounts the request to place it in the available experimental hall No. 205. Overall collider size including the biological shield has to match the hall dimensions. On the other hand the existing extraction channel of the Nuclotron defines the convenient ways for the beam injection into both collider rings. Therefore both collider rings have racetrack shape: each of them consists of two arcs and two long straight sections. Two *interaction points* (IP) are located in the centers of these sections. The rings are arranged one upon the other that allows one designing the SC magnets having two apertures in one yoke.

A long luminosity life-time will be provided by application of an electron or stochastic cooling. Each arc includes two short straight sections. One or two of them can be used for the electron cooling system. To provide a round beams in the collision points the collider will be operated at equal tunes in horizontal and vertical directions.

One of the IPs will be used for experiments with light ions and polarized deuterons. For spin manipulation in the collision point Siberian snakes are foreseen.

The ring lattice consists of "usual" FODO cells. Necessary space for diagnostic and correction elements is foreseen. The short straight section in the arcs is aimed at injection systems layout. The regular structure in the arcs contains 24 dipoles and 36 quadrupoles. The insertion of strong focusing quadrupoles around both IPs forms there the *beta-functions of 0.5 m at zero dispersion.*

Ion beams injection is produced in the horizontal plane in the short straight sections of the arcs using septum kicker magnets (identical systems for each ring).

The rms bunch length σ_s is below 30 cm that requires to apply either very *high RF voltage*, or to have a high *harmonics number* of RF. The RF bucket length $l_{separatrix}$ of 2.4 m ($8{\cdot}\sigma_s$).Consequently, the RF harmonics number can be ≤ 105. In the NICA collider we propose to use the 90th harmonics. The distance between bunches has to be as long as

$$Lgap \geq 0.5 \cdot (\tau_{kick} + \tau_{jit})\ \beta c + 0.5 \cdot l_{sep} \approx 15\ m\ .$$

This corresponds to 15 bunches in the ring of the 224 m circumference.

Beam cooling methods must be applied at experiments energy in the collider to avoid a luminosity decrease due to an increase of the ion beam emittance and the bunch length effected by intrabeam scattering (IBS). The estimated IBS growth time values are of the order of 50 s at the ion energy of 3.5 GeV/u. It means a loosing of luminosity if IBS is not suppressed by a cooling, because time of complete refreshment of the beams in both rings is of 150 s (Fig. 3).

Both electron and stochastic cooling method are being considered for the collider. The first one can be designed and constructed using the existing experience of Fermilab group, which had constructed and commissioned successfully he cooler of 4.3 MeV electron energy (about 8 GeV/u ion energy)[14] . A lower electron energy, of 2.4 MeV, is required for the NICA. Nevertheless, some development will have to be done: insertion of solenoid magnetic field, application of different type of HV power supply, and so on. This work is planned to be fulfilled in collaboration with the groups from FZ Juelich and Budker INP aiming also to design electron coolers for COSY (FZ-Juelich) and HESR (FAIR, GSI).[15] .

Application of stochastic cooling is retarded by the lack of a sufficient experience of bunched beam cooling. Only experiment performed couple years ago at RHIC (BNL)[16] demonstrated stochastic cooling of bunched gold ion beam in longitudinal degree of freedom. On the other hand, "direct" estimates of stochastic cooler parameters for NICA have given unrealistic value of band width W of the feed back system: above 8 GHz (due to huge peak current of circulating bunches). Some novel idea the\at allows to reduce W has been outspoken recently[17] , however it needs of a serious study.

The magnetic system of the collider will have to be operated at dipole field up to 4.5 T with very long ramping time (Table 1). Twin bore structural dipoles and quadrupoles with Cosθ type of superconducting coils are proposed (Fig. 11) As prototypes for these magnets can be used "the UNK magnets"[18] . However, the dipoles of NICA will have to be curved that requires an R&D as well.

6. Conclusion

The NICA project is dedicated to design and construction of a novel accelerator complex having unique parameters. The project realization will open possibilities of experimental studies in very wide range of high energy and relativistic nuclear physics.

References

1. *Design and Construction of Nuclotron-based Ion Collider fAcility (NICA), Conceptual Design Report*, Editors: I. Meshkov, A. Sidorin, 2008.
2. A. N. Sisakian, A. S. Sorin et al., These proceedings.
3. V. D. Kekelidze, A. N. Sisakian, A. S. Sorin et al., in *The MultiPupose Detector – MPD*, These proceedings.
4. A. N. Govorov et al., in *Proc. of the Conf. "LINAC'96"*, 394–396.
5. E. D. Donets, in *Physics and Technology of Ion Sources*, Ed. by I. G. Brown, Wiley &Sons, NY, 245 (1989).
6. E. D. Donets et al, Patent RU2067784, Bull. *Izobretenia*, Moscow, **28** (1996).
7. E. D. Donets et al., *Phys. Scripta T* **80**, 500 (1999); E. D. Donets, *Rev. Sci. Instrum.* **71**, 810 (2000); E.D.Donets et al., ibid., 887.
8. E. D. Donets et al., *Rev. Sci. Instrum.* **71**, 896 (2000);; E. D. Donets et al., in *Proc. of EPAC'98, Stockholm*, **2**, 1409 (1998); K.-G. Rensfelt et al., *Proc. of EPAC'2000, Vienna*, 578 (2000); D. E. Donets et al., *Proc. of EBIS/T'2000 Symposium, BNL, USA, November 2000, AIP Conf. Proc.* 572 (2001) 103.
9. V. A. Teplyakov, in *Proceedings of EPAC 2006, Edinburgh, Scotland THPPA03, Prize Presentation.*
10. N. I. Balalykin, V. G. Kadyshevsky, V. V. Kobets, I. N. Meshkov, I. A. Seleznev, G. D. Shirkov, A. N. Sissakian and E. M. Syresin, in *Proceedings of APAC 2004, Gyeongju, Korea.*
11. N. N. Agapov, A. M. Baldin, H. G. Khodzhibagiyan, A. D. Kovalenko, V. N. Kuzichev, V. A. Mikhailov, and A. A. Smirnov, *IEEE Trans. Appl. Supercond.* **10**, 280 (2000).
12. X. D. Yang, V. V. Parkhomchuk et al., in *Proc. of COOL'07 (2007)*, http://www.cool07.gsi.de

13. A. M. Baldin et al., *JINR preprint 9-11796, Dubna 1978.*
14. S. Nagaitsev, A. Bolshakov, D. Broemmelsiek, A. Burov et al, *AIP* **821**, 39 (2006).
15. V. Parkhomchuk et al., *AIP* **821**, 308 (2006).
16. J. M. Brennan and M. Blaskiewicz, in *Proc. of COOL 2007 Workshop, Bad Kreuznach, Germany, 2007.*
17. I. Meshkov, D. Moehl, T. Katayama, Stochastic cooling with the shifted PU electrodes, (2008) to be published.
18. A. I. Ageev, N. I. Andreev, V. I. Balbekov et al., in *Proceedings of EPAC, 1992.*

THE MULTIPURPOSE DETECTOR (MPD)

V. D. KEKELIDZE, A. N. SISSAKIAN, A. S. SORIN

Joint Institute for Nuclear Research,
Joliot Curie st., 6, Dubna, Moscow region, 141980 Russia
E-mail: kekel@sunse.jinr.ru ,
E-mail: sisakian@jinr.ru ,
E-mail: sorin@theor.jinr.ru

for the MPD Collaboration

A conceptual design of MultiPurpose Detector (MPD) is proposed to study the hot and dense baryonic matter in collisions of heavy ions over atomic mass range A = 1–238 at a center-of-mass energy up to $\sqrt{s_{NN}} = 9$ AGeV. The MPD experiment is foreseen to be carried out at the future JINR accelerator complex for heavy ions – the Nuclotron Based Ion Collider fAcility which is designed to reach required parameters with an average luminosity of $L = 10^{27} cm^{-2} s^{-1}$.

1. Introduction

Investigation of the hot and dense baryonic matter provides information on the in-medium properties of hadrons and nuclear matter equation of state, and allows to search for possible manifestation of deconfinement and/or chiral symmetry restoration, phase transition, mixed phase and critical end-point, and to shed light on the evolution of the Early Universe and formation of neutron stars [1–4] . It is a challenging task of modern physics. A study of hadronic matter was strongly inspired by the proposed hypothesis that the quarks should poses an additional quantum number (the color) [5–7] .

A new accelerator complex - the Nuclotron-based Ion Collider fAcility (NICA)[9] , will be constructed at Joint Institute for Nuclear Research (JINR) aiming the investigation of hot and dense baryonic matter. It has to provide collisions of heavy ions over atomic mass range A = 1–238 at a centre-of-mass energy up to $\sqrt{s_{NN}} = 9\ AGeV$ and average luminosity of $L = 27 cm^{-2} s^{-1}$. Two interaction points are foreseen at the NICA collider for mounting of two detectors operating simultaneously. One of these detectors - the MultiPurpose Detector (MPD), is optimized for the study of heavy-ion collisions and search for the phase transition, mixed phase and the critical end-point [8] .

2. Physics goals

The high-density nuclear matter can be created in a transient reaction volume in relativistic heavy ion collisions. In these collisions a large fraction of the beam energy is spent for production of hadrons including excited resonances whose properties may be modified by the surrounding hot and dense medium. At very high temperatures or densities, the hadron mixture melts and their constituents, quarks and gluons, form a new phase of matter, the quark-gluon plasma. The different phases of strongly interacting matter are shown in the phase diagram in Fig.1.

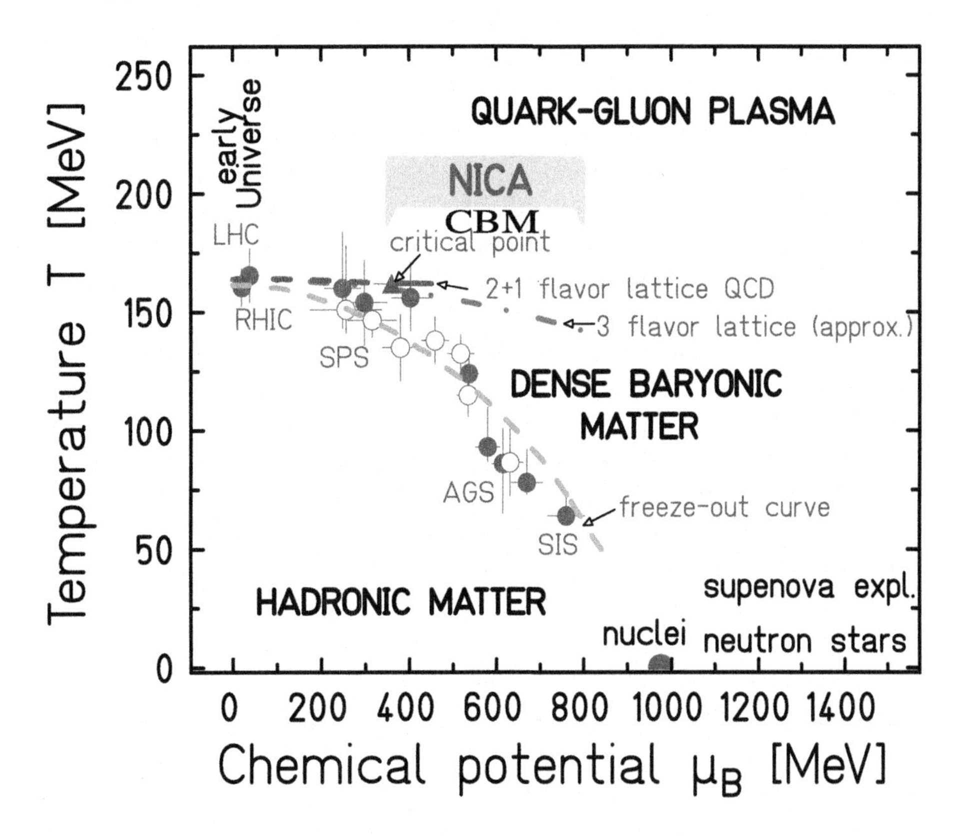

Fig. 1. The phase diagram of strongly interacting QCD matter. The symbols represent freeze-out points obtained with a statistical model analysis of particle ratios measured at the mid rapidity (open circles) and in the 4π acceptance (filled ones) in cental heavy-ion collisions[10] . The dashed curve is the chemical freeze-out calculation at the kinetic energy per baryon ratio $E/N_B = 1$ GeV[11] . The phase boundary is shown for the $N_f = 2 + 1$ lattice QCD calculation with a critical point (triangle) at $T = (162 \pm 2)$ MeV and $\mu_B = (360 \pm 40)$ MeV[12] . The dot-dashed line is a parabola with the slope corresponding to the lattice predictions of $d^2T/d\mu_B^2$ of the transition line at $\mu_B = 0$.

As it is seen from this phase diagram, several heavy-ion experiments at CERN-SPS, BNL-RHIC and CERN-LHC probe the region of high temperatures and low net baryon densities where circumstantial evidence has been obtained for a new phase of matter existing above a temperature of $T_E \sim 160 - 170\ MeV$. In the other region of the phase diagram, at lower temperature and moderate baryonic density, the GSI-SIS experiments definitely show no hint of a phase transition but certainly point to the in–medium modification effects. At very high densities and very low temperatures the matter is deconfined and, as predicted, correlated quark-antiquark pairs form a color superconductive phase. Such phase may be created in the interior of neutron stars.

One of the most interesting region of the phase diagram is the intermediate one, where the hadronic system enters a new phase at a beam energy of $\approx 30AGeV$ in a fixed target experiment. The fascinating particularity of this energy range is the critical end point located according to recent lattice QCD calculations at $T_E = (162 \pm 2)$ MeV and $\mu_E = (360 \pm 40)$ MeV[12] with model predictions covering the area of $T_E \approx 50 - 170$ MeV and $\mu_E \approx 200 - 1400$ MeV[13] .

The phase diagram translates into a visible pattern the properties of strong interactions and their underlying theory, Quantum Chromo Dynamics (QCD). In particular, such fundamental QCD phenomena as confinement and broken chiral symmetry, whose quantitative understanding is still lacking, are a challenge for future heavy-ion research.

The major goal of the MPD experiment is the study of in-medium properties of hadrons and the nuclear matter equation of state, including a search for possible signals of deconfinement and/or chiral symmetry restoration, phase transitions, and the QCD critical endpoint in the region of the collider energy $\sqrt{s_{NN}} = 4-9\ AGeV$. Due to the high complexity of this task and large uncertainty in the predicted signals, an accurate scaling of the considered phase diagram domain in collision energy, impact parameter and system size is utterly needed as well as the uniform acceptance for observables over the whole energy range of interest. The collider mode used in the MICA/MPD project naturally satisfies this demand.

3. Observables and Requirements for the Detector

The MPD envisaged experimental programme includes simultaneous measurements of the observables which are presumably sensitive to high density effects and phase transitions[4] . The observables measured on event-by-event basis are particle yields, their phase-space distributions, correlations and fluctuations. Different particle species probe different stages of the nucleus-nucleus interaction due to their difference in masses, energies, and interaction cross sections.

The gold–gold inelastic collisions in the energy range from $\sqrt{s_{NN}} = 3\ AGeV$ to 9 $AGeV$ were simulated by means of UrQMD (version 1.3) code with strong decays of all resonances taken into account. Fig. 2 and Fig. 3 represent the obtained charged particles multiplicity and pseudorapidity distributions, respectively.

The most promising observables which could serve as sensitive diagnostic probes in the considered energy range are the strangeness–to–entropy ratio which can be simulated by the kaon–to–pion and/or Lambda–to–pion ratios, multiplicity fluctuations, HBT correlations, and collective flows of identified hadrons. A study of these observables will allow one to extract information on the nuclear equation of state for baryonic matter at high densities. Prior to earching for these effects, one should estimate the kinematical characteristics of the particles produced and then consider their peculiarities and trace their energy dependence.

The event–by–event multiplicity fluctuations are expected to be an indication of crossing the boundary of the first order phase transition where these fluctuations should be large.

The elliptic flow (v_2), which is defined as a coefficient of expansion of angular distribution of emitted particles

$$dN/d\phi \propto [1 + 2v_1 \cos(\phi) + 2v_2 \cos(2\phi)]$$

is expected to be sensitive to the early pressure gradients and therefore to the equation of state (EOS) of the fireball formed in the heavy-ion collisions. This angular distribution of the particles emitted is non-isotropic and v_2 is a measure of stretching. In principle, the elliptic flow of hadrons at low transverse momenta (p_t) can be related to the degree of thermalization, the viscosity, and the EOS of the matter produced. At NICA/MPD energies v_2 strongly depends on collision energy and changes its sign. These dependencies serve as an important constraint for discriminating between various EOSs for high density nuclear matter, and provide

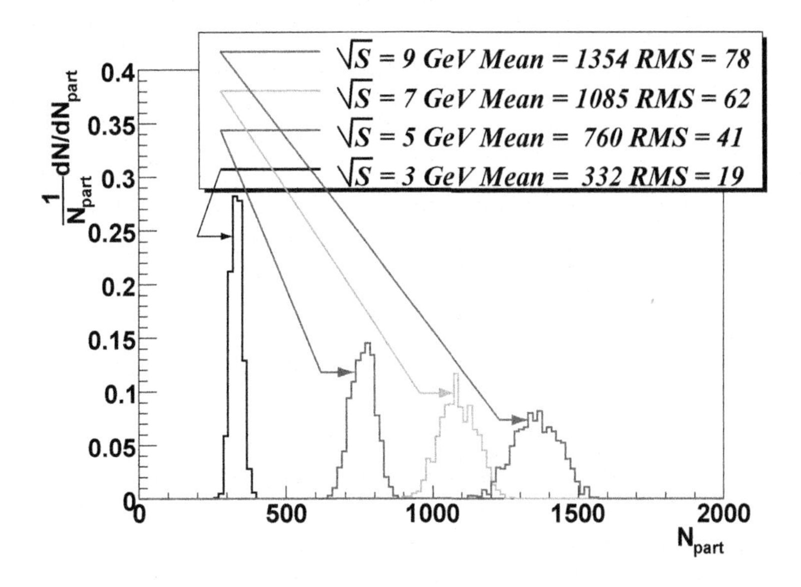

Fig. 2. Charged multiplicity distributions in central ($b = 0 - 3$ fm) collisions at different energies calculated by UrQMD.

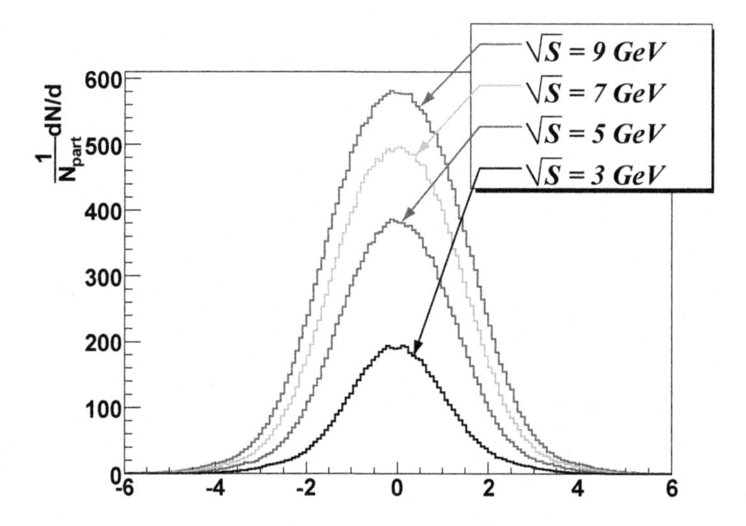

Fig. 3. Pseudorapidity distributions of charged particles in central ($b = 0 - 3$ fm) collisions calculated by UrQMD.

important insights on the interplay between collision geometry and the expansion dynamics.

Hadrons containing the strange quarks are of particular interest. At high densities and/or temperatures strange particles originating from the early stage of collision process should have rather high transverse momenta. The elliptic flow of the strange and non–strange hadronic matter is essential at midrapidity.

Information about the space-time structure of particle emission source (fireball) can be extracted by femtoscopy or HBT interferometry. UrQMD calculations of HBT-radii of particle emission source demonstrate essential diversity. Calculations of HBT-radii for NICA range of energy were performed with the fast hadron freeze-out MC generator (FASTMC)[14] . Particles and hadronic resonances are generated on the thermal chemical hypersurface. These radii are related to the correlation function as

$$C(q_0, q_S, q_L) = 1 + \lambda exp(-R_L^2 q_L^2 - R_0^2 q_0^2 - R_S^2 q_S^2 - 2R_{0L}^2 q_0 q_L)$$

Here, q_i and R_i are the components of the pair momentum difference $\mathbf{q}$ and the homogeneity length (HBT radii) in the i direction, respectively. The pre–factor λ is the incoherence parameter taking values between 0 (complete coherence) and 1 (complete incoherence). The term R_{0L}^2 vanishes at mid–rapidity for symmetric systems, and takes nonzero values at large rapidities.

The following processes are proposed to study at the first stage of the MPD experiment:

- multiplicity and spectral characteristics of the identified hadrons including strange particles, multi-strange baryons and antibaryons characterizing entropy production and system temperature at a final interaction stage
- event-by-event fluctuations in multiplicity, charges, transverse momenta and K/π ratios as a generic property of critical phenomena
- collective flows (directed, elliptic and higher ones) for observed hadrons including strange particles driven by the pressure in the system
- HBT interferometry of identified particles and particle correlations (femtoscopy)

The MPD detector should work at rate of about $6 \cdot 10^3$ interactions per second with multiplicities of up to $\approx$ 1400 charged particles per central Au+Au collision at the maximum energy.

4. General Design

The requirements to a detector at heavy ion collider pose significant challenges for the general design, choice of subdetectors and construction. The MPD will be mounted in a limited space along the beam line between the collider magnetic optics. This detector should meet the following requirements:

- compatibility for the event rate up to 10 kHz with a multiplicity up to 1500 charged particles;
- efficient reconstruction of events, primarily in the pseudorapidity region of $\mid \eta \mid \leq 1.0$, with high angular and momentum resolution for charged particles (from 100 MeV to 2000 MeV);
- reliable identification of charged particles for those events and possibility to detect $e^{\pm}$, γs and π^0s;
- capability to reconstruct momenta and identify tracks of charged particles in the region of $\mid \eta \mid \geq 1.0$;
- provide reliable information for "centrality" definition and corresponding trigger logic.

A basic concept of the MPD is represented by a barrel and two end-caps located inside the magnetic field of a large solenoid. The barrel part is a shall-like set of various detector systems surrounding the interaction point, and intended to reconstruct and identify both charged and neutral particles in the region of $\mid \eta \mid \leq 1.0$. Two symmetric end-cap parts are designed to reconstruct tracks and measure momenta of charged particles with higher pseudorapidity. A basic design of the MPD is shown in Fig.4.

The MPD general geometry is axially symmetric in accordance with the corresponding of events. The ion beams are going through the whole detector inside the beam pipe located along its z-axis with the central interaction point at $z = 0$ in the center of the detector. The interaction regions covers an interval of $-25 \leq z \leq 25$ cm.

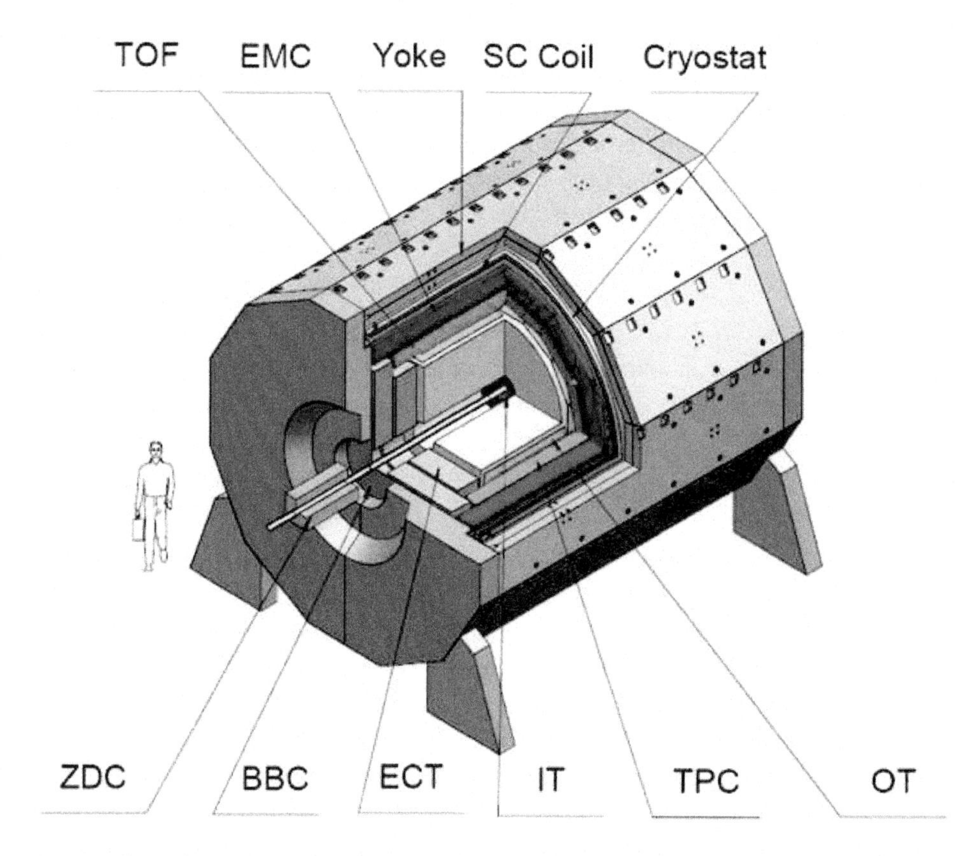

Fig. 4. General view of MPD detector with end doors retracted for access to inner detector components: superconductor solenoid (SC Coil) and magnet yoke, inner detector (IT), straw-tube tracker (ECT), time-projection chamber (TPC), straw-tube overlap tracker (OT), time-of-flight stop counters (TOF), electro-magnetic calorimeter (EMC), beam-beam counter (BBC), zero degree calorimeter (ZDC).

The barrel part consists of trackers and a particle identification system. The major tracker is a time projection chamber (TPC), which is supplemented by an inner tracker (IT) inside the TPC. The outer tracker (OT) surrounding the TPC is an optional one. Using both sub-detectors (IT and OT) improve the precision of charged particle tracks reconstruction and determination of their momenta.

The time of flight (TOF) system is used for the charged track identification. It includes fast start detectors located close to the z-axis, and RPC chambers surrounding the OT.

The electromagnetic calorimeter (EMC) is foreseen for gamma and electron/positron detection as well as for π^0 reconstruction.

The EMC is located within the barrel structure surrounding the barrel system of RPC's.

Two EndCap Trackers (ECT) are located at both sides of the TPC along the z-axis. The current design includes two wheel-like tracking detectors based on straw-tubes. The ECT is designed so as to provide more information (hit numbers) for tracks reconstruction at smaller radii where the TPC reconstruction ability degrades. Thus a supplementary principle is fulfilled for the tracking of charged particles with high pseudorapidoty ($| \eta | \geq 1.0$).

The particles emitted in the very forward / backward directions are detected by zero degree calorimeters (ZDC), which are used for centrality definition, event-by-event fluctuation measurements and trigger signals formation.

Two beam-beam counter systems (BBC) located symmetrically along the z-axis provide information for centrality definition, L0 trigger and additional information for the interaction point reconstruction.

The detector geometry is optimized with due regard to some constrains and compromises. The length along the z-axis is limited by 800 cm available between the collider optics. The radial dimensions are chosen as a compromise between the requirements of TOF and track reconstruction precision on the one hand, and limitation imposed by the cost of the magnet creating the homogeneous magnetic field the other. The volume of ~ 200 *cm* radius inside the magnet available for mounting of subdetectors, was determined as an optimal one. This provides sufficient performance for track reconstructions and particle identification. The chosen dimension allows to construct a superconducting solenoidal magnet at a reasonable cost.

5. Magnet Coil and Flux Return

The MPD magnet is a 0.5 T superconducting solenoid located inside a dodecahedral iron yoke for flux return and shaping of the magnetic field ot required homogeneity. The magnet inner radius ($\approx 200cm$) and its length ($\approx 540cm$ with a cryostat) are optimized with regards to the other detector subsystems. In general, this magnet is similar to many detector magnets used in high energy physics. The magnet system consists of a solenoid, a flux return and supporting systems including cryogenics. The magnet yoke provides confinement of an inverse magnetic flux and contributes to shaping of a magnetic field to the required homogeneity in the TPC region ($B_r/B_z \leq 0.001$).

A highly developed at JINR and well tested in practice in the Nuclotron magnets, the tubular superconducting cable cooled from within two-phase helium is chosen as a basis element for the MPD superconducting solenoid. The wending machine for manufacturing such cable is functioning properly and needs only minor modernizing. demanding only inappreciable modernizing. Using of such cable eliminates the need to manufacture the high-precision aluminum cylinder for bandaging and cooling the superconducting winding. The JINR pilot plant has all the necessary equipments to manufacture the other cryostat systems as well.

6. TPC for Central Tracking

The TPC is an ideal tracker working well in a high multiplicity environment. It is the main tracking detector of MPD central barrel and together with the ITS and TOF has to provide charged particle momentum measurement, particle identification and vertex determination with sufficient momentum resolution and two track separation. All these requirements need to be fulfilled at the NICA design luminosity which corresponds to an interaction rate of $< 10kHz$. The produced particle multiplicity is < 1500. TPC has to perform track finding and momentum measurement in the transversal momentum region $0.1\ GeV/c < p_t < 3\ GeV/c$ and for $|\eta| < 1.0$. The TPC should provide tracking efficiency over 90%, momentum resolution better than 2.5%, and two track separation $\approx 1cm$.

TPC will be cylindrical in shape with an outer radius of 110 cm, an inner one of 20 cm, and 300 cm in length with two 150 cm long drift areas along the axi separated by a central high voltage electrode plane.

The charged particles which cross the TPC create electrons by ionization. The electrons drift in the electric field to the Readout Chambers. Amplification factor of these chambers is $\approx 10^4$. The crucial parameters that determine track resolution are the diffusion constants and total drift length.

Mechanical structure cosists of outer vessels, end flanges, and inner vessels mado of carbon fiber composite and honeycomb (NOMEX).

The wire Readout Chamber sectors (30° in azimuth) with total of 24 chambers are mounted on each face of the cylinder, on the end-cap wheels. The overall area to be covered is $\approx 8.0\ m^2$. The chambers will be conventional multiwire proportional chambers with cathode pad readout. To keep the occupancy as low as possible and to ensure the necessary dE/dx and position resolution, there will be about 80 000 readout pads with an area of $0.5\ cm^2$ near the inner radius and $1.0\ cm^2$ near the outer one. The Front-End Electronics for the TPC has to be highly integrated. It will consist of three basic units for each channel: a low-impedance charge-sensitive preamplifier/shaper, a commercial 10 MHz 10-bit ADC, an ASIC with a digital filter for tail cancelation, base-line subtraction and zero-suppression circuitry, and a multiple event buffer, all to be implemented in CMOS technology. The Front End Electronics may be the same as used in the ALICE TPC. The Nd-YAG (266 nm) laser will be used for testing and calibration. The uniformity of the transverse magnetic field should be $\approx 6 \times 10^{-4}$.

7. Inner Tracker (IT)

The Inner Tracker (IT) is a detector system situated in the very heart of the MPD with sensors working at the closest distance from the interaction beam-to-beam volume.

The IT task is to improve the track finding of the TPC especially for events with high multiplicity, and to find decay vertices of the secondary particles before they reach active volume of the TPC.

The IT is a conventional Silicon Tracking System (STS) exploiting a widely used technique of Silicon Strip Detectors (SSD). Similar to the ALICE IT, the STS has a barrel-like structure with multiple sensors creating surfaces of eight cylindrical layers located at the radii within 4 to 16 cm from the interaction point. The total sensitive area created by ≈ 720 modules is around 2.6 m^2. This covers the rapidity range $|\eta| \leq 1$ for all vertices at the geometric center of the ITS.

The requirement to register low momentum protons and fragments makes necessary to develop a system with minimum possible material budget. Here, the technological developments of the ALICE Collaboration must be exploited at full. All layers support and cooling parts are to be manufactured of ultra-light carbon fiber composite materials[15] with all bonding of electrical contacts done by the modern TAB technology compatible with the usage of ultra light cables manufactured out of aluminum polyimide[16] .

The average thickness of one layer including the cooling and electrical services is around 1% X_0 with another 1% X_0 needed for electrical shielding of the sensors[17] . The total material budget of the STS for particles scattered at 90° is $\approx 7\%$ X_0.

Contrary to the ALICE's IT, STS basic element is a module of two single-sided SSDs. Two detectors of a module form a network of strips with a 100 mrad stereo angle between the strips of each successive sensor. This angle offers an acceptable rate of ambiguities in track reconstruction while keeping the number of read-out channels as small as possible with resolution along the beam axis of around 0.5 mm. The 230 μm strip pitch and the 60 mm strip length define the system granularity. With this pitch the occupancy is expected to be less than 10% in the inner layer of the ITS. All the sensors are 300 μm thick to provide good signal-to-nose ratio. The read-out system is an analog one with a dynamic range no less than 12 MIP to supply dE/dx values for identification of protons and deuterons. The first candidate to be considered is the HAL25 chip developed for the ALICE IT[18,19] .

The spacial resolution of the STS is 500 μm in the z-axis and 200 μm in the r_ϕ. Two track resolutions are 600 μm and 300 μm in the z-axis and the r_ϕ respectively.

8. Time of Flight System (TOF)

The TOF system of MPD is the main detector for particles identification allowing pion/kaon separation in the momentum range 0-2.5 GeV/c and proton/kaon separation in the range 0–4.5 GeV/c. The TOF system consists of two subsystems. the first one includes two stations of scintillation counters situated around the beam pipe on both sides from the detector, and aimed to give the start signal and to provide measurement of the z-coordinate of the interaction point. The second subsystem, the barrel, gives the stop signal. A multigap RPC (MRPC) is chosen as a fast detector for time measurements in the barrel subsystem. This type of the detector proved to be reliable, rather simple in construction and not expensive. It is developed to be used in TOF systems of ALICE, CBM, STAR, PHENIX and other experiments. The total time resolution of the system ($T_{start} - T_{stop}$) must be better than 100 ps, the occupancy must be below 10–15%, and the system must be able to work in a magnetic field up to 0.5 T.

The TOF barrel covers the kinematical region of $-1 < \eta < 1$. It is 355 cm in length and 1.3 m in radius (see Fig.5).

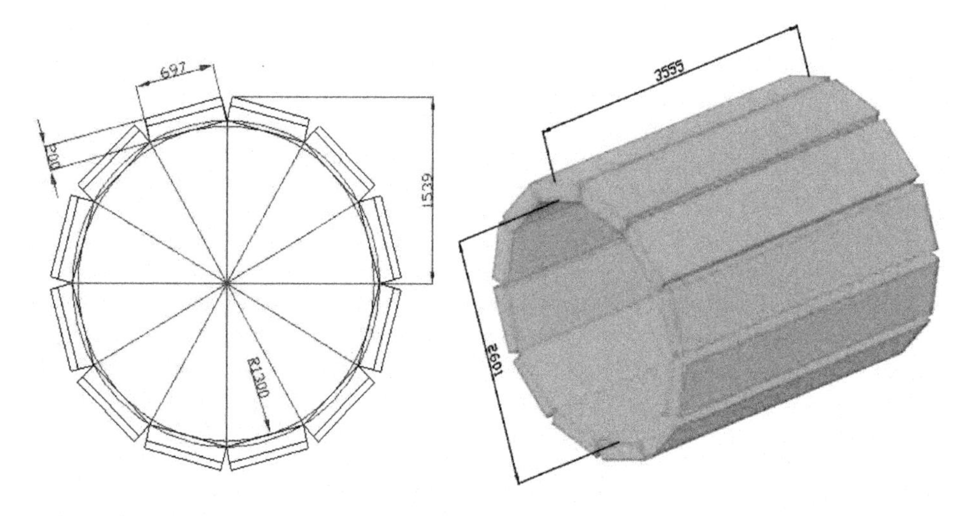

Fig. 5. Barrel of TOF.

The surface of the TOF barrel is $\approx 30\ m^2$. It is covered by 512 counters organized in 12 modules. There are 43 multigap RPCs in each module, placed perpendicular to the beam axis (see Fig.6). In order to exclude dead zones the counters in the module are placed with overlap. The area of one counter measures 7 cm$\times$ 67 cm, each of them has 64 readout pads $\approx 4\ cm^2$. The total number of readout channels is 32800. The MRPCs are placed in the module in such a way that readout pads are perpendicular to the line coming from the interaction point.

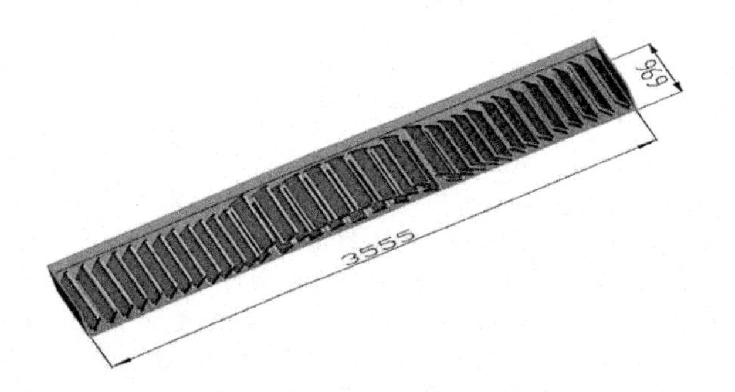

Fig. 6. View of the multigap RPCs distribution in the module box.

Each MRPC consists of a stack of 12 resistive plates of glass separated by 220 μm spacers creating 10 equal gas gaps. A 10 gap construction is chosen in order to avoid processing of small signals and to diminish multihit signals.

The density of charged particles created in central collisions is expected to be less than $0.003/\mathrm{cm}^2$ on the TOF barrel surface. The maximum rate per 1 cm^2 is $N = 10000Hz \times 0.003/cm^2 = 30Hz/cm^2$. So, the TOF system should provide reliable operation at particle flux up to 30 $\mathrm{Hz/cm}^2$.

The ALICE TOF collaboration[22] demonstrated that RPC made of commercial soda-lime glass with bulk resistivity of $\sim 10^{13}\ \Omega\cdot$ cm can operate at a flux in excess of 1 kHz with no degradation in the performance.

The start station should provide start signals for the barrel TOF system and for L0 trigger with an accuracy better than 50 ps. This allows one to measure the z-coordinate of the interaction point with an accuracy $\approx 1.5\ cm$. Cherenckov radiators coupled to photomultipliers may be used as basic elements of the Start detector.

9. Electromagnetic Calorimeter (EMC)

The main goal of the EMC is to detect $e\pm$ and γ and to measure energy. These measurements should provide precise determination of π^0/π^{+-} and γ/π^0 ratios, which may point to new phenomena in phase transition point. The measurement has to be done in the 50 MeV/c – 2 GeV/c momentum range. The π^0 yields will be extracted from the 2γs invariant mass spectrum, on top of a large combinatorial background. The rectangular 3×3 cm detector cells will guarantee the occupancy to be below 3% over the whole region. The energy resolution of the EMC is a critical point in the high combinatorial background environment (Fig.7, energy resolution below 2.5% being highly desirable. The study done by the COSY and KOPIO collaborations[20,21], shows such a possibility. Rejection of neutral hadrons can be achieved by a selection cut on the shower width, which works at all energies, and/or by a selection cut on time of flight. This may cover a relevant energy range. Finally, the EMC must be able to operate in a magnetic field of up to 0.5 T and be compact enough to be integrated into the MPD set-up.

Two solutions satisfy all listed requirements for the EMC. Lead tungstate ($PbWO_4$) is the most promising material for the photon detector. The light yield of lead tungstate crystals at room temperature is 5% of that of BGO, i.e. 40 - 65 photons/MeV for a 16 cm long crystal. However, the light yield is constantly being increased and the most recent crystals produced for CMS showed more than 80 photons/MeV. The temperature dependence offers the possibility to increase the intrinsic light yield of the crystal. At a working temperature of -25°C, the light yield will be enlarged by about a factor 2.5 and in addition the electronics noise of the photon detector will be reduced. Both effects will lead to improved energy resolution.

Another solution could be a Pb-scintillator electromagnetic calorimeter of shashlyk-type. Calorimeters of this type are relatively inexpensive, radiation hard, and have robust design. Their energy resolution can be as good as $\sim 3\%/\sqrt{E}$.

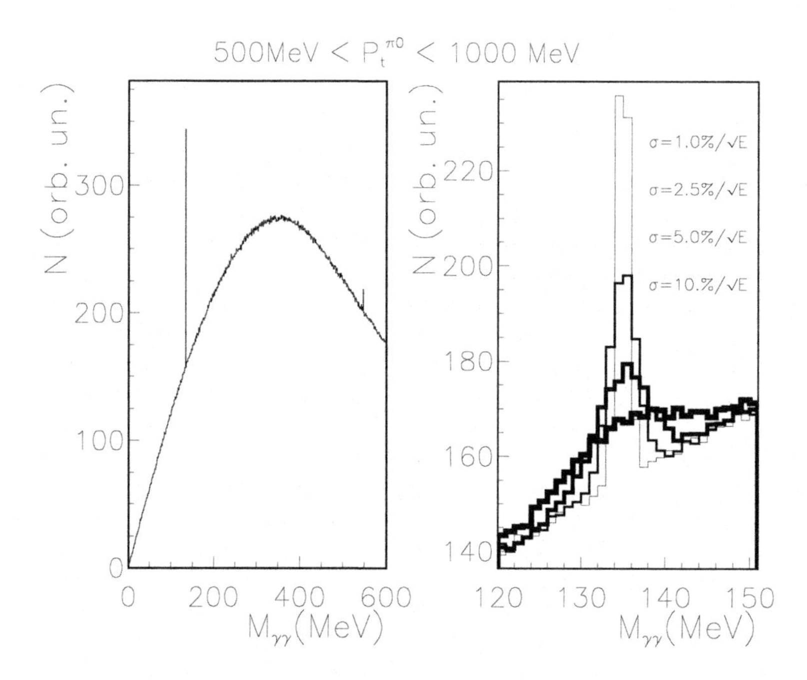

Fig. 7. π^0 reconstruction in the intermediate P_{π^0} transverse momentum regions for different photon detector energy resolution. Left plot – ideal detector. Direct Au-Au collision (4.5+4.5 AGev).

The basic element of the EMC is a 3×3×16(50) cm^3 module wrapped in Tyvek with light detectors glued onto the outer face of crystals. The 64 crystals arranged in a 8×8 square matrix form a 24×24×22(55) cm^3 module including light detectors and preamplifiers.

The proposed EMC includes 32640 detector cells in the barrel part with a total covered area of 28.6 m^2.

10. Endcap Tracker (ECT)

The Endcap Tracker (ECT) is proposed to extend the acceptance of the MPD to the pseudorapidity range of $1 < |\eta| < 2$ over the full azimuthal range on both faces of the detector. The increased acceptance improves the general event characterization in the MPD, in particular, event-by-event observables like $\langle p_T \rangle$, fluctuations of charged particle multiplicity and collective flow anisotropies. Furthermore, ECT will significantly enhance the triggering capabilities of MPD.

The ECT should provide the required spatial resolution for track reconstruction and operate successfully at high multiplicities. It will be located behind the both sides of the TPC occupying the region of (1550÷2330) mm along z-axis, and cover the θ angle from 10 to 43 degrees. Each of the Endcap detectors contains three (or two) wheel modules with an overall radius of ~1290 mm, 1730 mm and 1730 mm correspondingly. The first module consists of seven wheel submodules with the 960 mm long straws, second one contains six wheel submodules with the 1310 mm long straws and the third one contains three wheel submodules with the 1310 mm long straws. The thickness of the submodule with four straw layers is ~60 mm. Each wheel submodule consists of two main carbon-plastic rings with the holes for straws installation and for the outside gas-manifolds. The angle precision of the holes should be within 0.002°. The 4 mm in diameter reinforced straws will be positioned either radially or at a variable angle (tbd during R&D phase), as shown in Fig.8.

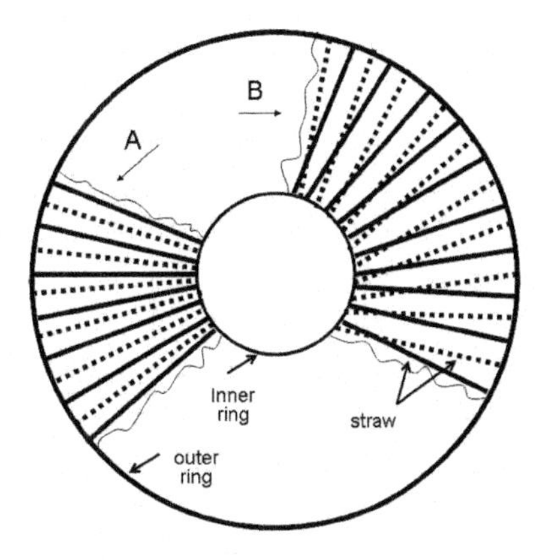

Fig. 8. Fragments of schematic designs for two subsequent straw layers : A) radial positioning; B) positioning at variable angle. The first straw layer is shown with solid lines, the dashed lines represent the second straw layer. Only one design will be selected for final EST setup.

The distance between the straw centers of two neighboring straws of the layer will be about 5.2 mm for the inner ring of the submodules. The ECT will be made of up to 64 straw layers per each side.

A primary track reconstruction for the proposed ECT was done for $\approx 200\ \mu m$ straw spacial resolution. The obtained relations hips of pseudorapidity to hit numbers, and occupancy to the inner radius are presented in Fig.9.

11. Zero Degree Calorimeter (ZDC)

In the nuclei-nuclei collisions the dense (and/or hot) hadronic matter is created in the more central collisions. To select central collisions and for triggering, energy of the forward emitted particles was proposed to be measured with a Zero Degree Calorimeter (ZDC)[30] ,[31] . In a simple geometrical picture, the forward-going matter reflects the degree of centrality of each event. In the more central collision many

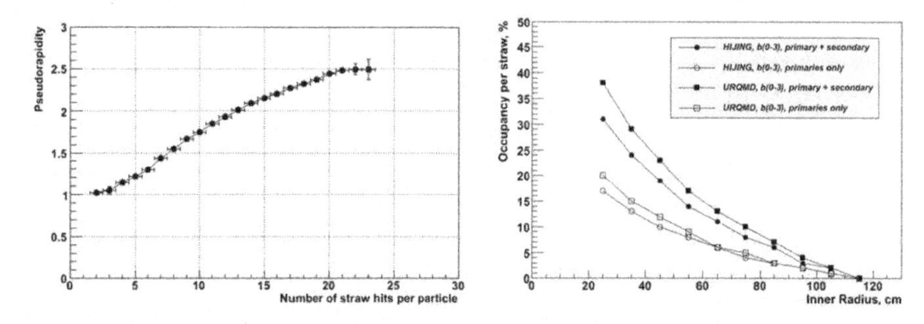

Fig. 9. Number of straw hits vs pseudorapidity (left) and occupancy per straw vs variable inner radius (right) of the first module of the Endcap Straw Tracker.

fewer fast particles enter a forward region of a small solid angle around the beam. In the peripheral collisions, a large fraction of the beam energy remains in the forward region.

In studying the event-by-event fluctuations, the experimental identification of the critical fluctuations requires a very accurate control on the fluctuations caused by the variation of the number of interacting nucleons due to event-by-event changes in the collision geometry. Therefore, the number of non-interacting nucleons from a projectile nucleus should be determined with the best possible precision. For this purpose the ZDC should have excellent energy resolution (better than $50\%/\sqrt{E}$) and transverse uniformity of this resolution ($< 5\%$). These parameters as well as the linearity in a wide energy range ($3 \div 1000$ GeV) and Gaussian shape of the detector signal are crucial in the appropriate description and subsequent correction of the physical parameters for the detector response.

In order to exclude fluctuations due to different number of neutral pions produced in the specific nucleus-nucleus collisions the ZDC should produce equal signals from hadronic and electromagnetic showers. This could be achieved by a full compensating modular lead-scintillator calorimeter.

It is proposed to place two ZDCs on each side of the detector along the beam axis at a distance of about 280 cm from the interaction point. Each ZDC will be assembled of 76 modules of hadron calorimeters. Each module of hadron calorimeter consists of 60 layers of lead-scintillator tile "sandwiches" with a sampling ratio of 4 : 1 (16 mm lead and 4 mm scintillator plates) to satisfy the compensation condition. The transverse dimension of the module is 10×10 cm^2. All 60 layers in each module are tied in one block 120 cm long (about 6 nuclear interaction lengths) by a 0.5 mm thick stainless steel tape.

Light readout is provided by the WLS-fibers embedded in the round grooves in scintillator plates that ensures high efficiency and uniformity of light collection over the scintillator tile within a very few percent. The WLS-fibers from each each group of 6 consecutive scintillator tiles are collected together and viewed by a single photo detector. The longitudinal segmentation in 10 sections ensures the uniformity of the light collection along the module as well as the rejection of secondary particles from interaction in the target. The 10 AMPDs (silicon photomultipliers, SiPMs) per module are placed at the rear side of the module together with the front-end-electronics (amplifiers). Such configuration ensures an easy access to the AMPDs.

12. Beam–Beam Counter (BBC)

The main role of the MPD Beam-Beam Counter (BBC) is to produce a signal for MPD Level-0 trigger. At low energies it is critical to have reliable minimum bias trigger sensitive as to the most central events so to the peripheral ones.

The proposed BBC consists of two scintillator annuli, installed around the beam pipe, on the two sides of the MPD magnet. The outer radius of the BBC is about 110 cm and the inner radius is just bigger than the 5 cm, that of beam pipe with a clearance between the BBC and beam pipe of 1 cm. This corresponds to a pseudo-rapidity range from 1.5 to 4.5 over full azimuth.

The BBC should be radiation hard and work in the magnetic field environment of $\approx 0.5T$. To satisfy these requirements, the BBC scintillators will be cut from the 1-cm thick radiation resistant Kuraray SCSN-81. The scintillation light produced within a tile is collected by four 0.83-mm diameter Y-11 doped optical fibers, inserted into grooves machined within the depth of the scintillator from both surfaces. The ends of the fibers are aluminized. The grooves ramp down from the scintillator surface and have fiber guides cut to trap the optical fibers. The fibers with a 3 cm minimum bend radius form a nearly circular loop within 2-mm from the isolation grooves to ensure response (light collection) uniformity volume.

The BBC plane is devided into regular hexagonal tiles defined by cutting 2-mm wide 5-mm deep optical isolation grooves in a Mercedes pattern on both sides of the scintillator, and then filling these grooves with the MgO2-loaded epoxy. Since grooves are cut on both sides, the optical isolation is complete. After machining, the sides of the hexagons are covered with white reflecting paint to trap scintillation light within each tile. Each scintillator surface is then covered with 1-mil thick aluminized mylar, taped to the painted scintillator edges. The reflectors are then covered by 10-mil thick black construction paper and black electrician's tape, to make the assembly light tight.

The pseudorapidity region covered by BBC ($1.5 < \eta < 4.5$) is out of the TPC acceptance and therefore will not introduce any trigger bias to the physics measurements in the midrapidity region ($-1.0 < \eta < +1.0$). A minimum bias trigger will require both BBCs to be fired to reduce beam gas background contamination. The further selection of vertex position using timing and amplitude information from BBCs will allow one to select clean minimum bias events.

According to HIJING simulations, the proposed BBC setup has rather high efficiency for triggering minimum bias events with $0 < b < 15.8$ fm. It is assumed that Zero Degree Calorimeter (ZDC) will provide trigger capabilities in the ultraperipheral region beyond this region(e.g. for $b > 15.8$).

13. Detector performance

The software for MPD experiment (mpdroot) is based on the FairRoot[32] framework providing a powerful tool for detector performance studies, development of pattern recognition algorithm for reconstruction and physics analysis. The framework is extensively used by CBM,[33] PANDA[34] and other experiments in GSI.

In the framework applied the detector response is simulated by a package currently based on Virtual Monte Carlo concept[35] allows performing simulation using Geant3, Geant4 or Fluka without changing the user code. The same framework is used for simulation and data analysis.

A useful advantage of the framework is that a new geometry reader was developed for it with the input in the form of TGeoVolumes (Root Geometry format). This reader is used by the PANDA collaboration to read the detector geometries which are converted from the Step file format (CAD system) to the Root format. After importing the geometry data, the database (MySQL) can be used to efficiently store the detector geometry, materials and parameters.

For realistic simulation of various physics processes an interface to the Monte Carlo event generators for nuclear collisions (UrQMD[36] and FastMC[14]) has been provided. Superposition of minimum-bias events can also be generated with the programme. The detector performances were obtained with a simpler version of the detector geometry. The more detailed detector description and careful cross checks with the above program will be made at the following stages of project preparation.

At the current stage of simulation the feasibility of event reconstruction with the TPC detector as the main tracking device was studied. The track finding efficiency was evaluated for all charged particles entering the TPC acceptance with the transverse momentum $p_t \geq 100$ MeV/c. Foe homogeneous magnetic field inside TPC a reasonable choice for pattern recognition to find hits, belonging to the track seems to be the conformal mapping method)[37] . As shown in Fig. 10, this method provides good efficiency of track finding in high multiplicity events. For track fitting the standard Kalman filter [38] was used and relative momentum errors $\Delta p/p \approx 1-2\%$ obtained for particles in the momentum range considered as shown in Fig. 10.

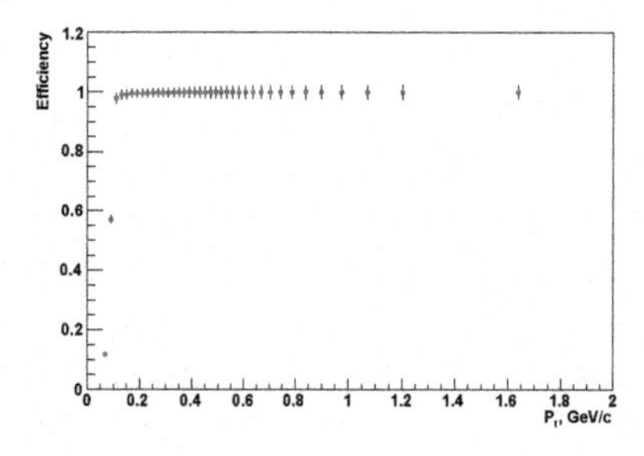

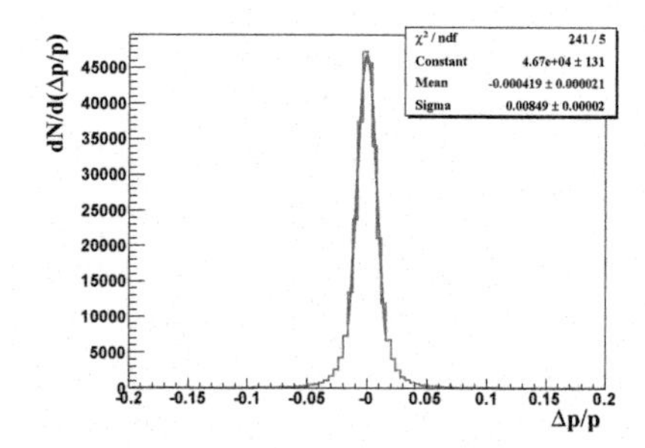

Fig. 10. Efficiency of track finding in TPC (left). Relative momentum errors for all particles (right).

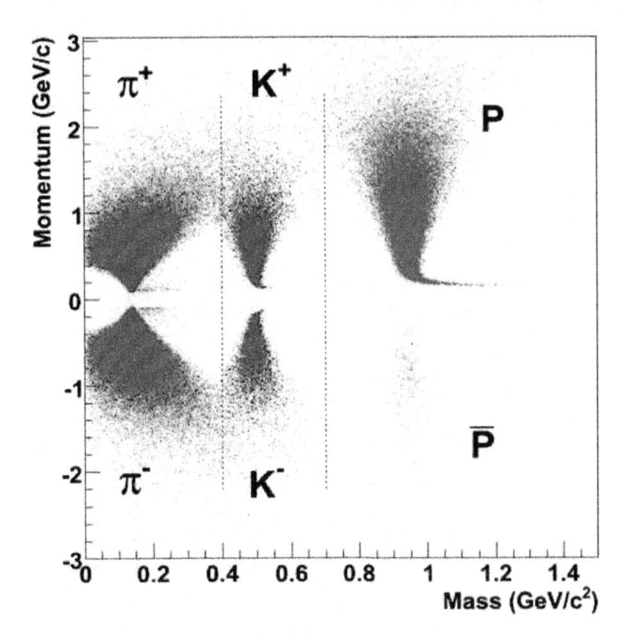

Fig. 11. Mass separation with TOF detector as a function of momentum (central collisions, primaries).

Fig.11 shows mass separation capabilities of TOF system.

In identifying K mesons, contamination level steeply increases at the momenta of particles exceeding 1.2 GeV/c. The efficiency of identification is estimated as $\approx 90\%$. The feasibility of the Λ-hyperon detection is confirmed for central Au+Au collisions as well .

14. Conclusion

The proposed conceptual design of the MPD detector is considered as a first step of project preparation. The collaboration started the corresponding R&D works and foresees its progress. The project will be supplemented by the missing parts related to DAQ, slow control, computing, engineering, various auxiliaries etc. The growing of the Collaboration could assure a further progress in the project execution.

15. Acknowledgements

The authors are grateful to M. N. Chubarov for the help in document preparation.

References

1. A. N. Sissakian, A. S. Sorin, M. K. Suleimanov, V. D. Toneev, and G. M. Zinovjev, Proceedings of the 8th International Workshop 'Relativistic Nuclear Physics: From Hundreds MeV to TeV', Dubna 2005, Russia, 306 (2006), nucl-th/0511018.

2. A. N. Sissakian, A. S. Sorin, M. K. Suleimanov, V. D. Toneev, and G. M. Zinovjev, *Bogoliubov Laboratory 50 years* (JINR, 2006), pp. 73 – 95, nucl-ex/0601034
3. A. N. Sissakian, A. S. Sorin, M. K. Suleimanov, V. D. Toneev, and G. M. Zinovjev, *Part. Nucl. Lett.* **5**, 1 (2008), nucl-ex/0511018.
4. A. N. Sissakian, A. S. Sorin, and V. D. Toneev, Proceedings of the 33rd International High Energy Physics conference, ICHEP'06, Moscow, Russia **1**, 421 (2006), nucl-th/0608032.
5. N. Bogolyubov, B Struminsky and A. Tavkhelidze, JINR Preprint D-1968, Dubna, 1965; A. Tavkhelidze, High Energy Physics and Elementary Particle, Vienna, 1965, p.753.
6. M.Y. Han, Y. Nambu, *Phys. Rev. B* **139**, 1005 (1965).
7. Y. Miyamoto, *Prog. Theor. Phys. Suppl. Extra.*, 1965, No.187.
8. A.N. Sissakian, A.S. Sorin, V.D. Kekelidze, *et al.*, *MPD Collaboration, The MultiPurpose Detector, Letter of Intent* (JINR, Dubna, 2006).
9. A. Sissakian *et al.*, *Design and Construction of Nuclotron-based Ion Collider fAcility (NICA), Conceptual design report* (JINR, Dubna, 2007).
10. A. Andronic, P. Braun-Munzinger, and J. Stachel, *Nucl. Phys. A* **772**, 167 (2006).
11. J. Cleymans and K. Redlich, *Phys. Rev. Lett.* **81**, 5284 (1998).
12. Z. Fodor and S. Katz, *JHEP* **404**, 50 (2004), hep-lat/0402006.
13. M. Stephanov, (to be published in PoS Lat2006) **024**, hep-lat/0701002.
14. N. S. Amelin *et al.*, Phys. Rev. **C74**, 064901 (2006), nucl-th/0608057.
15. G. A. Feofilov *et al.*, Inner Tracking System for ALICE: Conceptual Desing of Mechanics, Cooling and Alignement, in *Proceedings of the International Workshop on Advanced Materials for High Precision Detectors, Advance Proceedengs*, edited by C. H. B. Nicquevert, pp. 73–81, Geneve, 1994, CERN 94-07.
16. A. P. de Haas *et al.*, Aluminum microcable technology for ALICE silicon strip detector: status report, in *Proceedings of the Eighth Workshop on Electronics for LHC Experiments*, Colmar, France, 2002, CERN/LHCC 2002-034.
17. M. Bregant *et al.*, *Nucl. Intr. Meth. in Phys. Res. A* **569**, 29 (2006).
18. C. Hu-Guo *et al.*, The HAL25 front-end chip for ALICE silicon strip detectors, in *Proceedings of the Seventh Workshop on Electronics for LHC Experiments*, Stockholm, Sweden, 2001.
19. C. Hu-Guo *et al.*, Test and evaluation of HAL25: the ALICE SSD front-end chip, in *Proceedings of the Eighth Workshop on Electronics for LHC Experiments*, Colmar, France, 2002, CERN/LHCC 2002-03.
20. R. Novotny *et al.*, *IEEE Trans. Nucl. Sci.* **47**, 499 (2000).
21. G. Atoian *et al.*, *Nucl. Instrum. Meth. A* **531**, 467 (2004).
22. ALICE, (2000), CERN-LHCC-2000-012, ALICE TDR 8.
23. W. van Loo, *Phys. Stat. Sol.* **28**, 227 (1975).
24. W. Moses and S. E. Derenzo, *IEEE Trans. Nucl. Sci.* **36**, 173 (1989).
25. M. Kobayashi *et al.*, *Nucl. Instrum. Methods A* **333**, 429 (1993).
26. V. G. Baryshevsky *et al.*, *Nucl. Instrum. Methods A* **322**, 231 (1992).
27. P. Lecoq *et al.*, (1994), CMS TN/94-308.
28. Y. Borodenko *et al.*, in *Int. Conf. on Nuclear Tracking and Radiation Measurements*, Amsterdam, 1995.
29. M. N. Khachaturian, *Particles and Nuclei* **34**, 1316 (2003).
30. C. Adler *et al.*, *NIM A* **499**, 433 (2002).
31. NA49, H. Appelshauser *et al.*, *Eur. Phys. J. A* **2**, 383 (1998).
32. M. Al-Turany, D. Bertini, and I. Koenig, *CBM Simulation and Analysis Framework, GSI scientific report 2004* (GSI, 2004), FAIR-EXP-07.

33. http://www.gsi.de/fair/experiments/CBM.
34. http://www-panda.gsi.de.
35. http://alisoft.cern.ch/.
36. S. Bass *et al.*, *Prog. Part. Nucl. Phys.* **41**, 225 (1998), nucl-th/9803035.
37. P. Yepes, *NIM A* **380**, 582 (1996).
38. R. Fruhwirth, *NIM A* **262**, 444 (1987).

ON THE FIRST ORDER PHASE TRANSITIONS SIGNAL IN MULTIPLE PRODUCTION PROCESSES

[1,2] J. MANJAVIDZE and [1] A.SISSAKIAN

[1] *Joint Institute for Nuclear Research,*
Joliot Curie st., 6, Dubna, Moscow region, 141980 Russia
[2] *Andronikashvili Institute of Physics,*
Tamarashvili st., 6, Tbilisi, 380077 Georgia
E-mail: joseph@nu.jinr.ru
E-mail: sisakian@jinr.ru

We offer the parameter, interpreted as the "chemical potential", which is sensitive to the first order phase transition: it must decrease with number of evaporating (produced) particles (hadrons) if the (interacting hadron or/and QCD plasma) medium is boiling and it increase if no phase transition occur. The main part of the paper is devoted to the question: how one can measure the "chemical potential" in the hadron inelastic processes. Our definition of this parameter is quite general but assumes that the hadron multiplicity is sufficiently large. The simple transparent phenomenological lattice gas model is considered for sake of clarity only.

1. Introduction

Despite the fact that the first order phase transition in the ion collisions is widely discussed both from theoretical[1] and experimental[2] points of view the feeling of some dissatisfaction nevertheless remain. To all appearance the main problem consists in absence of the single-meaning directly measurable ("order") parameter(s) which may confirm this phenomenon in the high energy experiment. Our aim is to offer such parameter, explain its physical meaning and to show how it can be measured.

We guess that to observe first order phase transition it is necessary to consider very high multiplicity (VHM) processes. Then in this multiplicity region exist following parameter:

$$\mu(n,s) \simeq -\frac{T(n,s)}{n}\ln\sigma_n(s), \tag{1}$$

Here σ_n is the normalized to unite multiplicity distribution which can be considered in the VHM region as the "partition function" of the *equilibrium* system, see Appendix, and T is the mean energy of produced particles, i.e. T is associated with temperature. Continuing the analogy with thermodynamics one can say that $(-T\ln\sigma_n)/n$ is the Gibbs free energy per one particle. Then μ can be interpreted as the "chemical potential" measured with help of *observed* particles* in a free state.

We assume that the system obey the equilibrium condition, i.e. produced particles energy distribution can be described with high accuracy by Boltzmann exponent, or, it is the same, the inequality (15) must be satisfied. This assumption defines the "VHM region"[3] . It must be underlined that existence of the "good" parameter T does not assumes that the whole system is thermally equilibrium, i.e. the energy spectrum of unobserved particles may be arbitrary in our "inclusive" description.

The definition (1) is quiet general. It can be used both for hadron-hadron and ion-ion collisions, both for low and high energies. It is model free and operates only with "external" directly measurable parameters. The single indispensable condition: we work in the VHM region of observed particles. It is evident that such generality has definite defect: measuring μ one can not say something about details of the process.

This "defect" have following explanation. The point is that the classical theory of phase transitions have dealing immediately with the *internal* properties of media in which the transition occur. But in our, "S-matrix", case one can examine only the *external* response on the phase transition in the form of created mass-shell particles.

Continuing the analogy with the boiling, we are trying to define the boiling by the number of evaporating particles. The effect is evidently seen if the number of such particles is very large, i.e. in the VHM case. The "order parameter" is the work needed for one particle production, i.e. coincides with the "chemical potential". In the boiled "two-phase" region the media is unstable against "evaporation" of particles, i.e. the chemical potential must decrease with number of produced particles.

We offer quantitative answers on the following three question.

(A) *In what case one* may *observe first order phase transition.*
We will argue that observation of VHM states are necessary to find phase transition phenomenon. First of all the energy of produced particles are small in VHM case. This means that the kinetic degrees of freedom does not play essential role, i.e. they can not destroy, wipe out, the phase transition phenomenon. The second reason is connected with observation that in the VHM region one may use such equilibrium thermodynamics parameters as the temperature $T(n,s)$, chemical potential $\mu(n,s)$ and so on.

(B) *What we can measure.*
We will see that in VHM region exist the estimation (1) where all quantities in r.h.s. are measurables.

(C) *What kind effect one may expect.*
Chemical potential, $\mu(n,s)$, by definition is the work which is necessary to introduce, i.e. to produce, one particle into the system[4] . If the first order pase transition

*Notice that one may consider n as the multiplicity in the experimentally observable range of phase space.

occurs then $\mu(n,s)$ must decrease with n in the two-phase ("boiling") region. It is our general conclusion which will be explained using lattice gas model.

2. Definitions

2.1.

We will start from simple generalization of well known formulae. Let us consider the generating function

$$\rho(z,s) = \sum_{n=0}^{\infty} z^n \sigma_n(s). \tag{1}$$

For sake of simplicity σ_n is normalized so that

$$\rho(1,s) = 1. \tag{2}$$

One may use inverse Mellin transformation:

$$\sigma_n(s) = \frac{1}{2\pi i} \oint \frac{dz}{z^{n+1}} \rho(z,s) \tag{3}$$

to find σ_n if $\rho(z,s)$ is known. Noting that σ_n have sharp maximum over n near mean multiplicity $\bar{n}(s)$ one may calculate integral (3) by saddle point method. The equation:

$$n = \frac{\partial}{\partial \ln z} \ln \rho(z,s) \tag{4}$$

defines mostly essential value $z = z(n,s)$. Notice that $\sigma_n \equiv 0$ if the hadron multiplicity $n > n_{max} = \sqrt{s}/m$, where m is the characteristic mass of hadron. Production of identical particles is considered for sake of simplicity. Therefore, only $z < z_{max} = z(n_{max}, s)$ have the physical meaning.

One may write $\rho(z,s)$ in the form:

$$\rho(z,s) = \exp\left\{\sum_{l=0}^{\infty} z^l b_l(s)\right\}, \tag{5}$$

where the Mayer group coefficient b_l can be expressed through k-particle correlation function (binomial moments) $c_k(s)$:

$$b_l(s) = \sum_{k=l}^{\infty} \frac{(-1)^{(k-l)}}{l!(k-l)!} c_k(s). \tag{6}$$

Let us assume now that in the sum:

$$\ln \rho(z,s) = \sum_k \frac{(z-1)^k}{k!} c_k(s) \tag{7}$$

one may leave first term. Then it is easy to see that

$$z(n,s) = n/c_1(s), \ c_1(s) \equiv \bar{n}(s), \tag{8}$$

are essential and in the VHM region

$$\ln \sigma_n(s) = -n \ln \frac{n}{c_1(s)}(1 + O(1/\ln n)) = -n \ln z(n,s)(1 + O(1/\ln n)). \quad (9)$$

Therefore, in considered case with $c_k = 0,\ k > 1$, exist following asymptotic estimation for $n >> 1$:

$$\ln \sigma_n \simeq -n \ln z(n,s), \quad (10)$$

i.e. σ_n is defined in VHM region mainly by the solution of Eq.(4) and the correction can not change this conclusion. It will be shown that this kind of estimation is hold for arbitrary asymptotics of σ_n.

If we understand σ_n as the "partition function" in the VHM region then z is the *activity* usually introduced in statistical physics if the number of particles is not conserved. Correspondingly the chemical potential μ is defined trough z:

$$\mu = T \ln z. \quad (11)$$

Combining this definition with estimation (10) we define σ_n through μ. But, if this estimation does not depend from the asymptotics of σ_n, it can be used for definition of $\mu(n,s)$ through $\sigma_n(s)$ and $T(n,s)$. Just this idea is realized in (1).

2.2.

Now we will make the important step. To put in a good order our intuition it is useful to consider $\rho(z,s)$ as the *nontrivial* function of z. In statistical physics the thermodynamical limit is considered for this purpose. In our case the finiteness of energy $\sqrt{s}$ and of the hadron mass m put obstacles on this way since the system of produced particles necessarily belongs to the energy-momentum surface[†]. But we can continue theoretically σ_n to the range $n > n_{max}$ and consider $\rho(z,s)$ as the nontrivial function of z.

Let us consider the analog generating function which has the first $n < n_{max}$ coefficient of expansion over z equal to σ_n and higher coefficients for $n \geq n_{max}$ are deduced from continuation of theoretical value of σ_n to $n \geq n_{max}$. Then the inverse Mellin transformation (3) gives a good estimation of σ_n through this generating function if the fluctuations near $z(n,s)$ are Gaussian or, it is the same, if

$$\left. \frac{|2n - z^3 \partial^3 \ln \rho(z,s)/\partial z^3|}{|n + z^2 \partial^2 \ln \rho(z,s)/\partial z^2|^{3/2}} \right|_{z=z(n,s)} << 1. \quad (12)$$

Notice that if the estimation (10) is generally rightful then one can easily find that l.h.s. of (12) is $\sim 1/n^{1/2}$. Therefore, one may consider $\rho(z,s)$ as the nontrivial function of z considering $z(n,s) < z_{max}$ if $n_{max} >> n >> 1$.

[†]It must be noted that the canonical thermodynamic system belongs to the energy-momentum shell because of the energy exchange, i.e. interaction, with thermostat. The width of the shell is defined by the temperature. But in particle physics there is no thermostat and the physical system completely belongs to the energy momentum surface.

Then it is easily deduce that the asymptotics of $\sigma_n(s)$ is defined by the leftmost singularity, z_c, of in this way generalized function $\rho(z,s)$ since, as it follows from (4), the singularity "attracts" the solution $z(n,s)$ *in the VHM region.* In result, we may classify asymptotics of σ_n in the VHM region if (12) is hold.

Thus our problem is reduced to the definition of possible location of leftmost singularity of $\rho(z,s)$ over $z > 0$. It must be stressed that the character of singularity is not important for definition of $\mu(n,s)$ in the VHM region at least with $O(1/\ln n)$ accuracy. One may consider only three possibility: at $n \to \infty$

(A) $z(n,s) \to z_c = 1$,

(B) $z(n,s) \to z_c, \quad 1 < z_c < \infty$,

(C) $z(n,s) \to z_c = \infty$.

Other possibilities are nonphysical or extremely rear. Correspondingly one may consider only three type of asymptotics in the VHM region:

(A) $\sigma_n > O(e^{-n})$. We will see that in this case the isotropic momentum distribution must be observed, i.e. the energy, ε, distribution in this case is Boltzmann-like, $\sim e^{-\beta\epsilon}$;

(B) $\sigma_n = O(e^{-n})$. Such asymptotics is typical for hard processes with large transverse momenta, like for jets[5] ;

(C) $\sigma_n < O(e^{-n})$. This asymptotic behavior is typical for multiperipheral-like kinematics, where the longitudinal momenta of produced particles are noticeably higher than the transverse ones[6] .

We are forced to assume that the energy is sufficiently large, i.e. z_{max} is sufficiently close to z_c. In opposite case the singularity would not be "seen" on experiment.

Our aim is to give physical interpretation of this three asymptotics. The idea, as it follows from previous discussion, is simple: one must explain the nature of singularity z_c. It must be noted at the same time that in the equilibrium thermodynamics exist only two possibility, (A) and (C)[13] and just the case (A) corresponds to the first order phase transition.

Summarizing the results we conclude: *if the energy is sufficiently large, i.e. if z_{max} is sufficiently close to z_c, if the multiplicity is sufficiently large, so that (15) is satisfied and $z(n,s)$ can be sufficiently close to z_c, then one may have confident answer on the question: exist or not first order phase transition in hadron collisions.*

It must be noted here that the heavy ion collisions are the most candidates since z_c is easer distinguishable in this case.

2.3.

The temperature T is the another problem. The temperature is introduced usually using Kubo-Martin-Schwinger (KMS) periodic boundary conditions[7] . But this way assumes from the very beginning that the system (a) is equilibrium[8] and (b) is surrounded by thermostat through which the temperature is determined. The first condition (a) we take as the simplification which gives the equilibrium state where the time ordering in the particle production process is not important and therefore the time may be excluded from consideration.

The second one (b) is the problem since there is no thermostat in particle physics. For this reason we introduce the temperature as the Lagrange multiplier $\beta = 1/T$ of energy conservation law[3] . In such approach the condition that the system is in equilibrium with thermostat replaced by the condition that the fluctuations in vicinity of β are Gaussian.

The interesting for us $\rho(z, s)$ we define through inverse Laplace transform of $\rho(z, \beta)$:

$$\rho(z, s) = \int \frac{d\beta}{2\pi i \sqrt{s}} e^{\beta\sqrt{s}} \rho(z, \beta). \tag{13}$$

It is known[8] that if the interaction radii is finite then the equation (of state):

$$\sqrt{s} = -\frac{\partial}{\partial\beta}\rho(z, \beta) \tag{14}$$

have real positive solution $\beta(n, s)$ at $z = z(n, s)$. We will assume that the fluctuations near $\beta(n, s)$ are Gaussian. This means that the inequality[3] :

$$\left. \frac{|\partial^3 \ln \rho(z, \beta)/\partial\beta^3|}{|\partial^2 \ln \rho(z, \beta)/\partial\beta^2|^{3/2}} \right|_{z=z(n,s),\beta=\beta(n,s)} << 1 \tag{15}$$

is satisfied. Therefore, we prepare the formalism to find thermodynamic description of the processes of particle production assuming that this S-matrix conditions of equilibrium (12) and (15) are hold[‡].

We want to underline that our thermal equilibrium condition (15) have absolute meaning: if it is not satisfied then $\beta(n, s)$ loses every sense since the expansion in vicinity of $\beta(n, s)$ leads to the asymptotic series. In this case only the dynamical description of S-matrix can be used.

It is not hard to see[3] that

$$\frac{\partial^l}{\partial\beta^l} \ln R(z, \beta)|_{z=z(n,s),\beta=\beta(n,s)} =< \prod_{i=1}^{l} (\epsilon_i - < \epsilon >) >_{n,s} \tag{16}$$

is the l-point energy correlator, where $< ... >_{n,s}$ means averaging over all events with given multiplicity and energy. Therefore (15) means "relaxation of l-point correlations", $l > 2$, measured in units of the dispersion of energy fluctuations, $l = 2$. One can note here the difference of our definition of thermal equilibrium from thermodynamical one[9] .

[‡]Introduction of $\beta(n, s)$ allows to describe the system of large number of degrees of freedom in terms of single parameter, i.e. it is nothing but the useful trick. It is no way for this reason to identify entirely $1/\beta(n, s)$ with thermodynamic temperature where it has self-contained physical sense.

2.4.

Let us consider now the estimation (1). It follows from (3) that, up to the preexponential factor,

$$\ln \sigma(n, s) \approx -n \ln z(n, s) + \ln \rho(z(n, s), s). \tag{17}$$

We want to show that, in a vide range of n from VHM area,

$$n \ln z(n, s) \gtrsim \ln \rho(z(n, s), s). \tag{18}$$

Let as consider now the mostly characteristic examples.

(A) *Singularity at* $z = 1$.

This case will be considered in Sec.3. In the used lattice gas approximation $\ln z(n) \sim n^{-5}$ and

$$\ln \sigma_n \approx n^{-4} = n \ln z(n)(1 + O(1/n)). \tag{19}$$

(B) *Singularity at* $z = 1 + 1/\bar{n}_j(s) < \infty$:

$\ln \rho(z, s) = -\gamma \ln(1 - \bar{n}_j(s)(z - 1))$. In this case $z(n) = 1 + 1/\bar{n}_j(s) - 1/n\bar{n}_j(s)$ and

$$\ln \sigma_n \approx -n/\bar{n}_j(s) + \gamma \ln n =$$

$$= -n/\bar{n}_j(s)(1 + O(\ln n/n)). \tag{20}$$

(C) *Singularity at* $z = \infty$: $\ln \rho(z, s) = c_k(s)(z - 1)^k,\ k \geq 1$.

In this case $z(n) = (n/kc_k)^{1/k} >> 1$ and

$$\ln \sigma_n \approx -n \ln z(n)(1 + O(1/\ln n)). \tag{21}$$

One can conclude:

(i) The definition (1) in the VHM region is rightful since the correction falls down with n. On this stage we can give only the estimation of correction but (1) gives the correct n dependence.

(ii) Activity $z(n, s)$ tends to z_c from the right in the case (A) and from the left if we have the case(B) or (C).

(iii) The accuracy of estimation of the "chemical potential" (1) increase from (C) to (A).

3. Ising model: phase transition

The physical meaning of singularity over z[10] may be illustrated by following simple model. As was mentioned above the singularity at $z = 1$ is interpreted as the first order phase transition. Therefore, let us assume[11] that β is so large that interacting particles strike together into clusters (drops). Then the Mayer's group coefficient for the cluster from l particle is

$$b_l(\beta) \sim e^{-\beta\tau l^{(d-1)/d}},$$

where $\tau l^{(d-1)/d}$, $l >> 1$, is the surface tension energy, d is the dimension. Therefore, if $d > 1$ the series over l in (5) diverges at al $z > 1$. At the same time, the sum (5) converge for $z < 1$.

We consider following analog model to describe condensation phenomenon in the particle production processes. Let us cover the space around interaction point by the net assuming that if the particle hit the knot we have (-1) and $(+1)$ in opposite case. This "lattice gas" model[4] has a good description in terms of Ising model[12] . We may regulate number of down oriented "spins", i.e. the number of produced particles, by external magnetic field $\mathcal{H}$. Therefore the "activity" $z = e^{-\beta\mathcal{H}}$, i.e. $-\mathcal{H}$ is the "chemical potential"[13] .

Calculation of the partition function means summation over all spin configurations with constraint $\sigma^2 = 1$. Here the ergodic hypothesis is used. It allows to exclude the time from consideration.

To have the continuum model we may spread normally the δ-function of this constraint[16] :

$$\delta(\sigma^2 - 1) \sim e^{-(1-\sigma^2)^2/\Delta}.$$

Therefore, the grand partition function of the model in the continuum limit looks as follows[13,17] :

$$\rho(\beta, z) = \int D\sigma e^{-S(\sigma,\mathcal{H})}, \tag{1}$$

where the action

$$S(\sigma, \mathcal{H}) = \int d^3x \left\{ \frac{1}{2}(\nabla\sigma)^2 - \varepsilon\sigma^2 + \alpha\sigma^4 - \lambda\sigma \right\}. \tag{2}$$

The structure of contributions in (1) essentially depends on the sign of constant ε, see Fig.1 where the case $\varepsilon > 0$ is shown. Following notations was used:

$$\varepsilon \sim (1 - \frac{\beta_c}{\beta}), \ \alpha \sim \frac{\beta_c}{\beta} > 0, \ \lambda \sim (\beta\beta_c)^{1/2}\mathcal{H}, \tag{3}$$

where $1/\beta_c$ is the phase transition temperature. Phase transition takes place if $\beta > \beta_c$ $(T < T_C)$, i.e. we will consider in present section $\varepsilon > 0$. In this case the mean spin $< \sigma > \neq 0$. We will assume that $\beta >> \beta_c$ since in this case the fluctuation around $< \sigma >$ are small and calculations in this case became simpler. Considered model describes decay of unstable vacuum[18] .

The singularity over $\mathcal{H}$ appears by following reason. At $\mathcal{H} = 0$ the potential

$$v = -\varepsilon\sigma^2 + \alpha\sigma^4 \tag{4}$$

have two degenerate minima at $\sigma = \pm\sqrt{\varepsilon/2\alpha}$. The external field $\mathcal{H} < 0$ we destroy this degeneracy. But in this case the system in the right-hand minimum (with the up-oriented spins) becomes unstable.

The branch point in the complex plane corresponds to this instability. The discontinuity gives[?] :

$$\rho(\beta, z) = \frac{a_1(\beta)}{\mathcal{H}^4} e^{-a_2(\beta)/\mathcal{H}^2}, \tag{5}$$

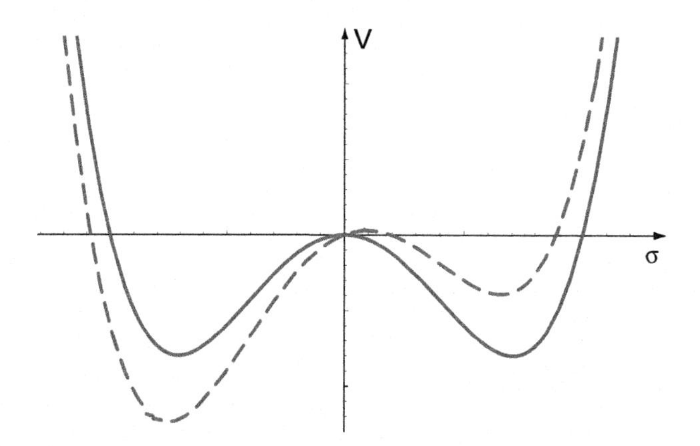

Fig. 1. Solid line: undisturbed by $\mathcal{H}$ potential and dotted line includes $\mathcal{H}$.

where $(\beta > \beta_c)$

$$a_1(\beta) = \frac{\pi^2}{2}\left(\frac{8\beta R^4}{9\beta_c A}\right)^{7/2}\left(\frac{1-\beta_c/\beta}{R^2}\right)^{3/4}\frac{R^4}{(\beta\beta_c)^2},$$

$$a_2(\beta) = \frac{8\pi}{81\sqrt{2}}\frac{\beta}{\beta_c}\left(1-\frac{\beta_c}{\beta}\right)^{7/2}\frac{R^4}{(A\beta_c^2)^2}. \qquad (6)$$

It must be noted that the eqs. (4) and (14) have only one solution:

$$\beta(n,s) \to \beta_c, \quad \ln z(n,s) \equiv l(n,s) \to 0.$$

at increasing n. This means that the singularities at $T = T_c$ and $z = 1$ attracts the solution:

$$l(n,\beta) \sim n^{-1/3}(\beta-\beta_c)^{7/6}, \quad \beta(n,s) = \beta_c(1+\gamma/n^4), \qquad (7)$$

where γ is the positive constant.

In result,

$$\ln\rho_n(s) \sim -n^{2/3}(1/n^{2/3})^7 \sim -1/n^4 \quad (\sim -n\ln z(n,s)) \qquad (8)$$

decrees with n and the chemical potential

$$\mu(n,s) \sim \frac{T_c}{n^5}(1+\gamma/n^4)^{-1} \qquad (9)$$

also decrees with n.

Some comments will be useful to this Section:

1. One may note that σ_n is defined by the discontinuity the the branch point in complex plane of $\ln z$ and decay of the meta-stable states does not play any role.

2. It follows from (7) that, at fixed β,

$$\ln z_c \sim (1/n)^{1/3} << 1. \qquad (10)$$

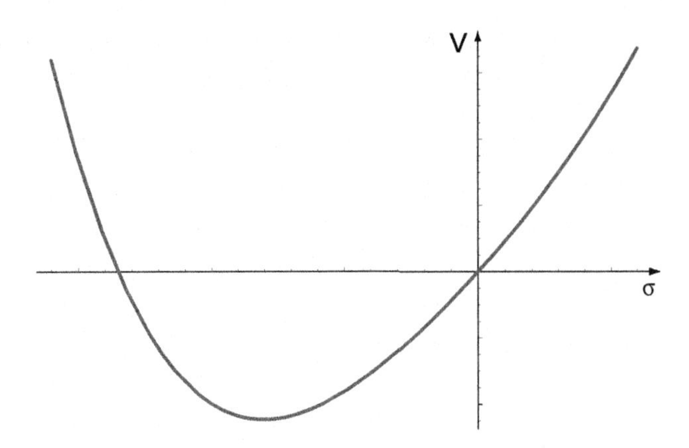

Fig. 2. Stable ground state disturbed by $\mathcal{H}$.

This means that for large n our calculations are correct. At the same time, in VHM region z near unite is essential and it *decrees* with n.

3. The work which is needed for production of one particle is $\sim \ln z(n,s)/\beta(n,s)$. Therefore production of large number of particles needs less work per one particle.

This conclusion have simple physical meaning (see beginning of present section). Let us consider decay of nonstable phase. The decay happens through production of clusters (domains with down oriented spins). The volume energy of cluster is $\propto R_0^3$, where R_0 is radius of cluster. It burst the dimension of cluster. If $R_0 < R_c$, where R_c is critical dimension of cluster, then the formation of such cluster is improbable. But if $R_0 > R_c$ then the probability grows with radii of cluster. Latter explains why the chemical potential falls down with multiplicity.

4. In the VHM region the temperature, $T(n.s)$, tends to its critical value, T_c, and slowly depends on n.

4. Ising model: stable minimum

Let as consider the system with stable vacuum, $\beta < \beta_c$ ($T > T_c$) in (3). In this case, see Fig.2, the potential $v(\sigma)$ has unique minimum at $\sigma = 0$. Switching on external field $\mathcal{H}$ the minimum move and the average spin appears, $\bar\sigma(\mathcal{H}) \neq 0$. One can find it from the equation:

$$-\triangle\sigma + 2\varepsilon\sigma + 4\alpha\sigma^3 = \lambda,\ \varepsilon > 0.$$

Having $\bar\sigma \neq 0$ we must expand the integral (1) near $\bar\sigma$:

$$\rho(\beta, z) = e^{\int dx\lambda\bar\sigma} e^{-W(\bar\sigma)}, \tag{1}$$

where $W(\bar\sigma)$ expandable over $\bar\sigma$:

$$W(\bar\sigma) = \sum_{l=1}^{\infty} \frac{1}{l} \int \prod_{i=1}^{l} \{dx_i \bar\sigma(x_i; \mathcal{H})\} B_l(x_1, ..., x_l), \tag{2}$$

where B_l is the l-point one particle irreducible vertex function. In another wards, B_l play the role of virial coefficient. Comparing (2) with (5) one may consider $\bar{\sigma}$ as the affective activity of group of l particles.

The representation (2) is useful since in the VHM region the density of particles is large and the particles momentum is small. Then, remembering that the virial decomposition is equivalent of decomposition over specific volume, calculating B_l one may not go beyond the one-loop approximation, i.e. we may restricted by the semiclassical approximation.

Therefore, having large density one may neglect the spacial fluctuations. In this case the integral (1) is reduced down to the the usual Cauchy integral:

$$\rho(\beta, z) = \int_{-\infty}^{+\infty} d\sigma e^{-(\epsilon\sigma^2 + \alpha\sigma^4 + \lambda\sigma)}. \tag{3}$$

In the VHM region $\lambda \sim \mathcal{H} \sim \ln z >> 1$ is essential. It is easy to see that

$$\bar{\sigma} \simeq -(\lambda/4\alpha)^{1/3} \tag{4}$$

is the extremum. The estimation of integral near this $\bar{\sigma}$ looks as follows:

$$\rho(\beta, z) \propto \left\{ 12\alpha \left(\frac{\lambda}{4\alpha} \right)^{2/3} \right\}^{-1/2} e^{3\lambda^{4/3}/4(4\alpha)^{1/3}}. \tag{5}$$

This leads to increasing with n activity:

$$l(n, s) \sim n^{8/3} \tag{6}$$

and

$$\beta(n, s) \sim n^{2/3}. \tag{7}$$

In result,

$$\ln \rho_n(s) \sim -n^{11/3} \ (\sim -n \ln z(n, s)). \tag{8}$$

A few comments at the end of this section:

(i) Cross section falls dawn in considered case faster then $O(e^{-n})$. The estimation:

$$\rho_n(s) \sim e^{-n \ln \bar{z}(n,s)}$$

gives the right expression in the VHM region.

(ii) The chemical potential increase with n:

$$\mu(n, s) = -(T(n, s)/n) \ln \sigma_n(s) \sim n^2. \tag{9}$$

5. Conclusions

We may conclude that:

(i) We found definition of chemical potential (1). This important observable can be measured on the experiment directly where $T(n, s)$ is the mean energy of produced particles at given multiplicity n and energy $\sqrt{s}$.

((ii) Being in the VHM region one may consider that $\mu(n, s) = O(1/n)$ at comparatively high multiplicities and it rise, $\mu(n, s) = O(n)$ with rising multiplicity, Fig.3, at comparatively low multiplicities. The transition region is defines the critical temperature T_c. But it is quiet possible that the condition (15) allows to see only one branch of the curve shown on Fig.3.

iv) The simplest example of finite z_c presents the jet considered in the case (B), Fig.1. Hence case (C) has pure dynamical basis and can not be explained by equilibrium thermodynamics.

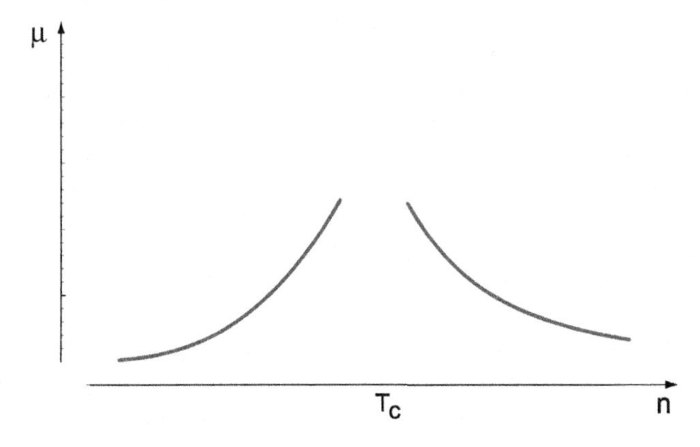

Fig. 3. Chemical potential $\mu(n, s)$ as the function of multiplicity n. T_c is the critical temperature. Breakthrough is the "two-phase" region.

Acknowledgements

We would like to thank participants of 7-th International Workshop on the "Very High Multiplicity Physics" (JINR, Dubna) for stimulating discussions.

References

1. M. Creutz, *Phys. Rev. D*,**15** 1128 (1977); M. Gazdzicki and M. I. Gorenstein, *Acta Physica Polonica B*, **30** 2705 (1999); A. N. Sissakian, A. S. Sorin, V. D. Toneev (Dubna, JINR) in *Proceedings of 33rd International Conference on High Energy Physics (ICHEP 06), Moscow, Russia, 26 Jul - 2 Aug 2006* e-Print: nucl-th/0608032
2. *BNL Report, Hunting the Quark Gluon Plasma*, BNL-73847-2005; C. Alt et al. arXiv: nucl-ex/0710.0118
3. J. Manjavidze and A. Sissakian, *Phys. Rep.*, **346** 1 (2001).

4. A. Isikhara, *Statistical physics*, Mir, Moscow (1973).
5. I. C. Taylor, *Phys.Lett. B*, **73** 85 (1978); A. J. MacFarlane and C. Woo, *Nucl.Phys. B*, **77** 91 (1974).
6. E. Kuraev, L. Lipatov and V. Fadin, *Sov. Phys. JETP*, **44** 443 (1976); *Zh. Eksp. Teor. Fiz.*, **71** 840 (1976); L. Lipatov, *Sov. J. Nucl. Phys.*, **20** 94 (1975); V. Gribov and L. Lipatov, *Sov. J. Nucl. Phys.*, **15** 438, 675 (1972); G. Altarelli and G. Parisi, *Nucl. Phys. B*, **126** 298 (1977); I. V. Andreev, *Chromodynamics and Hard Processes at High Energies* (Nauka, Moscow, 1981).
7. A. J. Niemi and G. Semenoff, *Ann.Phys. (NY)*, **152** 105 (1984); N. P. Landsman and Ch. G. vanWeert, *Phys.Rep.*, **145** 141 (1987).
8. M. Martin and J. Schwinger, *Phys.Rev.*, **115** 342 (1959).
9. N. N. Bogolyubov, *Studies in Statistical Mechanics,* (North-Holland Publ. Co., Amsterdam, 1962).
10. T. D. Lee and C. N. Yang, *Phys.Rev.*, **87** 404,410 (1952); S. Katsura, *Adv. Phys.*, **12** 391 (1963); H. N. Y. Temperley, *Proc.Phys.Soc. (London) A*, **67** 233 (1954).
11. A. F. Andreev, *Sov.Phys. JETP* **45** 2064 (1963).
12. C. F. Newell and E. W. Montroll, *Rev.Mod.Phys.*, **25** 353 (1953); M. E. Fisher, *Rep. Prog.Phys.*, **30** 731 (1967).
13. J. S. Langer, *Ann.Phys.*, **41** 108 (1967).
14. J. Schwinger, *J. Math. Phys. A*, **9** 2363 (1994); L. Keldysh, *Sov.Phys. JETP*, **20** 1018 (1964); P. M. Bakshi and K. T. Mahanthappa, *J. Math. Phys.*, **4** 1 (1961); *ibid.*, **4** 12 (1961).
15. J. Manjavidze and A. Sissakian, *Field-Theoretical Description of Restricted by Constrains Energy Dissipation Processes*, Preprint JINR P2-2001-117 (2001).
16. K. Wilson and J. Kogut, *Sov.Phys. NFF*, **5** (1975).
17. M. Kac, *Statistical mechanics of some one-dimensional systems*, Stanford Pub. (1962).
18. M. V. Voloshin, I. Yu. Kobzarev and L. B. Okun, *Sov.Phys. Nucl.Phys.*, **20** 1229 (1974); S. Coleman, *Phys.Rev. D*, **15** 2929 (1977); H. J. Katz, *Phys.Rev. D*, **17** 1056 (1978).
19. J. Manjavidze and A. Sissakian, *JINR Rap. Comm.*, **5(31)** 5 (1988).

BOSE-EINSTEIN CONDENSATION IN VERY HIGH MULTIPLICITY EVENTS

V. V. BEGUN[1] and M. I. GORENSTEIN[1,2]

[1] *Bogolyubov Institute for Theoretical Physics, Kiev, Ukraine*
[2] *Frankfurt Institute for Advanced Studies, Frankfurt, Germany*

We consider the Bose-Einstein condensation (BEC) in a relativistic pion gas. The thermodynamic limit when the system volume V goes to infinity as well as the role of finite size effects are studied. At $V \to \infty$ the scaled variance for particle number fluctuations, $\omega = \langle \Delta N^2 \rangle / \langle N \rangle$, converges to finite values in the normal phase above the BEC temperature, $T > T_C$. It diverges as $\omega \propto V^{1/3}$ at the BEC line $T = T_C$, and $\omega \propto V$ at $T < T_C$ in a phase with the BE condensate. At high collision energy, in the samples of events with a fixed number of all pions, N, one may observe a prominent signal. When N increases the scaled variances for particle number fluctuations of both neutral and charged pions increase dramatically in the vicinity of the Bose-Einstein condensation line. As an example, the estimates are presented for $p + p$ collisions at the beam energy of 70 GeV.

Keywords: Bose–Einstein condensation; High pion multiplicities; Finite size effects.

1. Introduction

The phenomenon of Bose-Einstein condensation (BEC) was predicted long time ago[1] . Tremendous efforts were required to confirm BEC experimentally. The atomic gases are transformed into a liquid or solid before reaching the BEC point. The only way to avoid this is to consider extremely low densities. At these conditions the thermal equilibrium in the atomic gas is reached much faster than the chemical equilibrium. The life time of the *metastable* gas phase is stretched to seconds or minutes. This is enough to observe the BEC signatures. Small density leads, however, to small temperature of BEC. Only in 1995 two experimental groups succeeded to create the 'genuine' BE condensate by using new developments in cooling and trapping techniques[2] . Leaders of these two groups, Cornell, Wieman, and Ketterle, won the 2001 Nobel Prize for this achievement.

Pions are spin-zero bosons. They are the lightest hadrons copiously produced in high energy collisions. There were several suggestions to search for the Bose-Einstein condensation (BEC) of π-mesons (see, e.g., Ref.[3]). However, no clear experimental signals were found up to now. Most of previous proposals of the pion BEC signals were based on an increase of the pion momentum spectra in the low (transverse) momentum region. These signals appear to be rather weak and they are contaminated by resonance decays to pions. In our recent papers[4,5] it was suggested that the pion number fluctuations strongly increase and may give a prominent signal of approaching the BEC. This can be achieved by selecting special samples of collision events with high pion multiplicities.

In the present talk we discuss the dependence of different physical quantities on the system volume V and, in particular, their behavior at $V \to \infty$. As any other phase transition, the BEC phase transition has a mathematical meaning in the thermodynamic limit (TL) $V \to \infty$. To define rigorously this limit one needs to start with a finite volume system. Besides, the finite size effects are important for an experimental search of the pion BEC fluctuation effects proposed in Ref.[4,5] . The size of the pion number fluctuations in the region of the BEC is restricted by a finite system volume V. To be definite and taking in mind the physical applications, we consider the ideal pion gas. However, the obtained results are more general and can be also applied to other Bose gases.

For more details see Ref.[6] . In Section II we consider the BEC in the TL. Section III presents a systematic study of the finite size effects for average quantities and for particle number fluctuations. In Section IV we discuss the finite size restrictions on the proposed fluctuation signals of the BEC in high energy collisions with large pion multiplicities. A summary, presented in Section V.

2. BEC in Thermodynamic Limit

We consider the relativistic ideal gas of pions. The occupation numbers, $n_{\mathbf{p},\mathbf{j}}$, of single quantum states, labelled by 3-momenta $\mathbf{p}$, are equal to $n_{\mathbf{p},j} = 0, 1, \ldots, \infty$, where index j enumerates 3 isospin pion states, π^+, π^-, and π^0. The grand canonical ensemble (GCE) average values, fluctuations, and correlations are the following[7] ,

$$\langle n_{\mathbf{p},j} \rangle = \frac{1}{\exp[(\sqrt{\mathbf{p}^2 + m^2} - \mu_j)/T] - 1} \tag{1}$$

$$\langle (\Delta n_{\mathbf{p},j})^2 \rangle \equiv \langle (n_{\mathbf{p},j} - \langle n_{\mathbf{p},j} \rangle)^2 \rangle = \langle n_{\mathbf{p},j} \rangle (1 + \langle n_{\mathbf{p},j} \rangle) \equiv v^2_{\mathbf{p},j},$$
$$\langle \Delta n_{\mathbf{p},j} \Delta n_{\mathbf{k},i} \rangle = v^2_{\mathbf{p},j} \delta_{\mathbf{p}\ \mathbf{k}} \delta_{ij} \tag{2}$$

where the relativistic energy of one-particle states is taken as $\epsilon_{\mathbf{p}} = (\mathbf{p^2} + m^2)^{1/2}$ with $m \cong 140$ MeV being the pion mass (we neglect a small difference between the masses of charged and neutral pions), T is the system temperature, and chemical potentials are $\mu_+ = \mu + \mu_Q$, $\mu_- = \mu - \mu_Q$, and $\mu_0 = \mu$, for π^+, π^-, and π^0, respectively. In Eq. (1) there are two chemical potentials: μ_Q regulates an electric charge, and μ a total number of pions. We follow our proposal of Refs.[4,5] and discuss a pion system with $\mu_Q = 0$ [a]. This corresponds to zero electric charge Q which is

[a] The BEC in the relativistic gas of 'positive' and 'negative' particles at $\mu_Q \to m$ has been discussed in Refs.[8–12]

defined by a difference between the number of π^+ and π^- mesons, $Q = N_+ - N_- = 0$. The total pion number density is equal to:

$$\begin{aligned} \rho(T,\mu) &\equiv \rho_+ + \rho_- + \rho_0 = \frac{\sum_{\mathbf{p},j}\langle n_{\mathbf{p},j}\rangle}{V} \\ &\cong \frac{3}{2\pi^2}\int_0^\infty \frac{p^2 dp}{\exp\left[\left(\sqrt{p^2+m^2} - \mu\right)/T\right] - 1} \\ &\equiv \rho^*(T,\mu) = \frac{3\,T\,m^2}{2\,\pi^2}\sum_{n=1}^{\infty}\frac{1}{n}\,K_2\,(n\,m/T)\,\exp(n\,\mu/T) \end{aligned} \tag{3}$$

where $\rho_j = \langle N_j\rangle/V$, with $j = +,-,0$, are the pion number densities, V is the system volume, and K_2 is the modified Hankel function. In the TL, i.e. for $V \to \infty$, the sum over momentum states is transformed into the momentum integral, $\sum_{\mathbf{p}} \dots = (V/2\pi^2)\int_0^\infty \dots p^2 dp$. The particle number density ρ depends on T and μ, and volume V does not enter in Eq. (3). This is only valid at $\mu < m$. The number of particles at zero momentum level is then finite, and its contribution to particle number density goes to zero in the TL. The inequality $\mu \leq m$ is a general restriction in the relativistic Bose gas, and $\mu = m$ corresponds to the BEC. The Eq. (3) gives the following relation between the BEC temperature T_C and total pion number density ρ[4] :

$$\rho = \frac{3\,T_C\,m^2}{2\pi^2}\sum_{n=1}^{\infty}\frac{1}{n}\,K_2\,(n\,m/T_C)\,\exp(n\,m/T_C) \tag{4}$$

A phase diagram of the ideal pion gas in the $\rho - T$ plane is presented in Fig. 1. The line of the BEC phase transition is defined by Eq. (4), and it is shown by the solid line in Fig. 1. In the non-relativistic limit, $T_C/m \ll 1$, using $K_2(x) \cong \sqrt{\pi/(2x)}\exp(-x)$ at $x \gg 1$, one finds from Eq. (4),

$$T_C \cong 2\pi\,[3\zeta(3/2)]^{-2/3}\,m^{-1}\,\rho^{2/3} \cong 1.592\,m^{-1}\,\rho^{2/3} \tag{5}$$

whereas in the ultra-relativistic limit[b], $T_C/m \gg 1$, one uses $K_2(x) \cong 2/x^2$ at $x \ll 1$, and Eq. (4) gives[4] ,

$$T_C \cong [3\zeta(3)/\pi^2]^{-1/3}\,\rho^{1/3} \cong 1.399\,\rho^{1/3} \tag{6}$$

In Eqs. (5-6), $\zeta(k) = \sum_{k=1}^{\infty} n^{-k}$ is the Riemann zeta function[15] , $\zeta(3/2) \cong 2.612$ and $\zeta(3) \cong 1.202$. The equation (5) corresponds to the well known non-relativistic result (see, e.g., Ref.[7]) with pion mass m and 'degeneracy factor' 3.

The particle number density is inversely proportional to the proper particle volume, $\rho \propto r^{-3}$. Then it follows from Eq. (5) for a ratio of the BEC temperature in

[b]The BE condensate formed in the ultra-relativistic regime has been considered in Refs.[13,14] as a *dark matter* candidate in cosmological models.

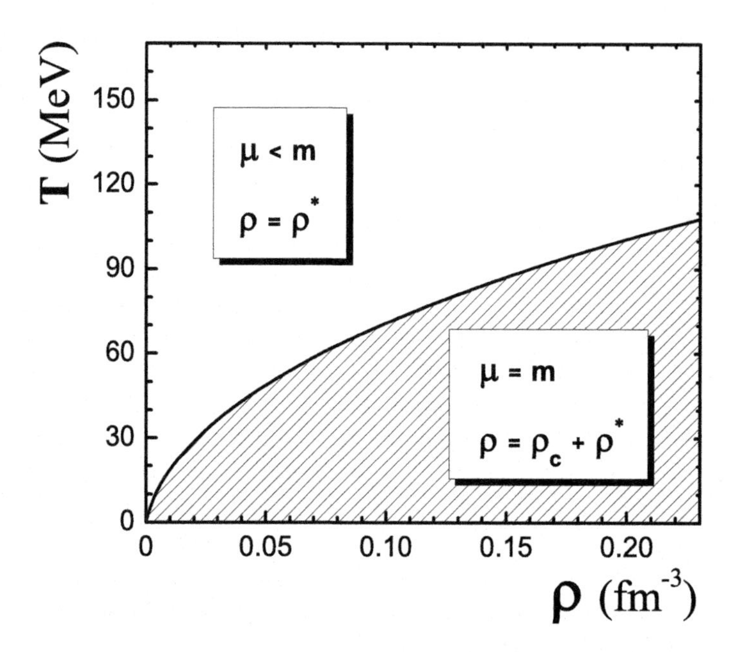

Fig. 1. The phase diagram of the relativistic ideal pion gas with zero electric charge density $\rho_Q \equiv \rho_+ - \rho_- = 0$. The solid line shows the relation $T = T_C$ (4) of the BEC. Above this line, $T > T_C$ and there is a normal phase described by Eq. (3). Under this line, $T < T_C$ and there is a phase with the BE condensate described by Eq. (8).

the atomic gases, $T_C(A)$, to that in the pion gas, $T_C(\pi)$, in non-relativistic approximation,

$$\frac{T_C(\pi)}{T_C(A)} \cong \frac{m_A}{m}\left(\frac{r_A}{r_\pi}\right)^2 \cong \frac{m_A}{m}\,10^{10} \tag{7}$$

where $r_A \cong 10^{-8}$ cm and $r_\pi \cong 10^{-13}$ cm are typical radiuses of atom and pion, respectively, and m_A is the mass of an atom. The Eq. (7) shows that $T_C(\pi) \gg T_C(A)$, and this happens due to $r_\pi \ll r_A$.

The equation (3) gives the total pion number density at $V \to \infty$ in the normal phase $T > T_C$ without BE condensate. At $T < T_C$ the total pion number density becomes a sum of two terms,

$$\rho = \rho_C + \rho^*(T, \mu = m) \tag{8}$$

The second term in the r.h.s. of Eq. (8) is given by Eq. (3). The BE condensate ρ_C defined by Eq. (8) corresponds to a macroscopic (proportional to V) number of particles at the lowest quantum level $p = 0$.

To obtain the asymptotic expansion of $\rho^*(T,\mu)$ given by Eq. (3) at $\mu \to m - 0$ we use the identity $[\exp(y) - 1]^{-1} = [\mathrm{cth}(y/2) - 1]/2$ and variable substitution, $p = \sqrt{2}m^{1/2}(m-\mu)^{1/2}x$, similar to those in a non-relativistic gas[16] . Then one finds,

$$\rho^*(T,m) \ - \ \rho^*(T,\mu) \ = \ \frac{3m^{3/2}}{\sqrt{2}\pi^2}(m-\mu)^{3/2}$$
$$\times \int_0^\infty x^2 dx\Big[\mathrm{cth}\frac{\sqrt{2m(m-\mu)x^2+m^2} \ - \ m}{2T} \ - \ \mathrm{cth}\frac{\sqrt{2m(m-\mu)x^2+m^2} \ - \ \mu}{2T}\Big]$$
$$\cong \ \frac{3m^{3/2}}{\sqrt{2}\pi^2}(m-\mu)^{3/2} \int_0^\infty x^2 dx\Big[\mathrm{cth}\frac{(m-\mu)x^2}{2T} \ - \mathrm{cth}\frac{(m-\mu)(x^2+1)}{2T}\Big]$$
$$\cong \ \frac{6Tm^{3/2}}{\sqrt{2}\pi^2}(m \ - \ \mu)^{1/2} \int_0^\infty x^2 dx\Big[\frac{1}{x^2} \ - \ \frac{1}{x^2+1}\Big] \ = \ \frac{3Tm^{3/2}}{\sqrt{2}\pi} \ (m \ - \ \mu)^{1/2} \quad (9)$$

At constant density, $\rho^*(T,\mu) = \rho^*(T = T_C, \mu = m)$, one finds in the TL at $T \to T_C + 0$,

$$\rho^*(T = T_C, \mu = m) \ = \ \rho^*(T,\mu) \ \cong \ \rho^*(T, \mu = m) - \frac{3Tm^{3/2}}{\sqrt{2}\pi}(m-\mu)^{1/2}$$
$$\cong \ \rho^*(T = T_C, \mu = m) \ + \ \frac{d\,\rho(T,\mu = m)}{dT}\Big|_{T=T_C} \cdot (T - T_C) \quad (10)$$
$$- \ \frac{3T_C \, m^{3/2}}{\sqrt{2}\pi} \ (m-\mu)^{1/2}$$

Using Eq. (10) one finds the function $\mu(T)$ at $T \to T_C + 0$,

$$m \ - \ \mu(T) \ \cong \frac{2\pi^2}{9T_C^2 \, m^3}\left[\frac{d\,\rho(T,\mu = m)}{dT}\Big|_{T=T_C}\right]^2 \cdot (T - T_C)^2 \quad (11)$$

In the non-relativistic limit $m/T \gg 1$ one finds, $\rho(T,\mu = m) \cong 3\zeta(3/2)\,[mT/(2\pi)]^{3/2}$, similar to Eq. (5). Using Eq. (11) one then obtains[16] ,

$$\frac{m \ - \ \mu(T)}{T_C} \ \cong \ \frac{9\zeta^2(3/2)}{16\pi} \ \cdot \ \left(\frac{T - T_C}{T_C}\right)^2 \quad (12)$$

In the ultra-relativistic limit, $T/m \gg 1$, one finds, $\rho(T,\mu = m) \cong 3\zeta(3) \ T^3/\pi^2$, similar to Eq. (6). This gives,

$$\frac{m \ - \ \mu(T)}{T_C} \ \cong \ \frac{18 \ \zeta^2(3)}{\pi^2} \ \left(\frac{T_C}{m}\right)^3 \ \cdot \ \left(\frac{T - T_C}{T_C}\right)^2 \quad (13)$$

Thus, $\mu(T) \to m$ and $d\mu/dT \to 0$ at $T \to T_C + 0$, both $\mu(T)$ and $d\mu/dT$ are continuous functions at $T = T_C$.

3. Finite Size Effects

3.1. *Chemical Potential at Finite Volume*

The standard introduction of ρ_C with Eq. (8) is rather formal. To have a more realistic picture, one needs to start with finite volume system and consider the limit $V \to \infty$ explicitly. The main problem is that the substitution, $\sum_{\mathbf{p}} \cdots \cong$

$(V/2\pi^2)\int_0^\infty \ldots p^2 dp$, becomes invalid below the BEC line. We consider separately the contribution to the total pion density from the two lower quantum states,

$$\rho \cong \frac{1}{V}\sum_{\mathbf{p},j}^{\infty}\langle n_{\mathbf{p},j}\rangle = \frac{3}{V}\frac{1}{\exp\left[(m-\mu)/T\right] - 1}$$
$$+\frac{3}{V}\frac{6}{\exp\left[\left(\sqrt{m^2+p_1^2}-\mu\right)/T\right]-1} \tag{14}$$
$$+\frac{3}{2\pi^2}\int_{p_1}^{\infty} p^2 dp\,\frac{1}{\exp\left[\left(\sqrt{m^2+p^2}-\mu\right)/T\right]-1}$$

The first term in the r.h.s. of Eq. (14) corresponds to the lowest momentum level $p = 0$, the second one to the first excited level $p_1 = 2\pi V^{-1/3}$ with the degeneracy factor 6, and the third term approximates the contribution from levels with $p > p_1 = 2\pi V^{-1/3}$. Note that this corresponds to free particles in a box with periodic boundary conditions (see, e.g., Ref[20]). At any finite V the equality $\mu = m$ is forbidden as it would lead to the infinite value of particle number density at $p = 0$ level.

At $T < T_C$ in the TL one expects a finite non-zero particle density ρ_C at the $p = 0$ level. This requires $(m-\mu)/T \equiv \delta \propto V^{-1}$ at $V \to \infty$. The particle number density at the $p = p_1$ level can be then estimated as,

$$\rho_1 = \frac{18\ V^{-1}}{\exp\left[\left(\sqrt{m^2+p_1^2}-\mu\right)/T\right]-1} \cong \frac{18\ V^{-1}}{\delta\ +\ p_1^2/(2mT)} \propto \frac{V^{-1}}{V^{-2/3}} = V^{-1/3} \tag{15}$$

and it goes to zero at $V \to \infty$. Thus, the second term in the r.h.s. of Eq. (14) can be neglected in the TL. One can also extend the lower limit of integration in the third term in the r.h.s. of Eq. (14) to $p = 0$, as the region $[0, p_1]$ contributes as $V^{-1/3} \to 0$ in the TL and can be safely neglected. Therefore, we consider the pion number density and energy density at large but finite V in the following form:

$$\rho \cong \frac{3}{V}\frac{1}{\exp\left[(m-\mu)/T\right] - 1} + \rho^*(T,\mu) \tag{16}$$
$$\varepsilon \cong \frac{3}{V}\frac{m}{\exp\left[(m-\mu)/T\right] - 1} + \varepsilon^*(T,\mu) \tag{17}$$

Thus, at large V, the zero momentum level defines completely the finite size effects of the pion system.

The behavior of $\rho^*(T,\mu)$ at $\mu \to m$ can be found from Eq. (9). At large V, Eq. (16) takes then the following form,

$$\rho \cong \frac{3}{V\,\delta} + \rho^*(T,\mu) \cong \frac{3}{V\,\delta} + \rho^*(T,\mu=m) - \frac{3}{\sqrt{2}\pi}(mT)^{3/2}\sqrt{\delta} \tag{18}$$

The Eq. (18) can be written as,

$$A\,\delta^{3/2}\;+\;B\,\delta\;-\;1\;=\;0 \tag{19}$$

where

$$A\;=\;\frac{V}{\sqrt{2}\pi}\;(mT)^{3/2}\;\equiv\;a(T)\;V \tag{20}$$

$$B\;=\;\frac{V}{3}\;[\;\rho\;-\;\rho^*(T,\mu=m)\;]\;\equiv\;b(T)\;V \tag{21}$$

The Eq. (19) for δ has two complex roots and one real root. An asymptotic behavior at $V \to \infty$ of the physical (real) root can be easily found. At $T < T_C$ it follows from Eq. (21) that $b(T) > 0$, and one finds from Eq. (19) at large V,

$$\delta\;\cong\;\frac{1}{b}\;V^{-1} \tag{22}$$

From Eq. (21) one finds that $b = 0$ at $T = T_C$. In this case, Eq. (19) gives,

$$\delta\;\cong\;\frac{1}{a^{2/3}}\;V^{-2/3} \tag{23}$$

The Eq. (19) can be also used at $T > T_C$, if T is close to T_C, thus, $\delta \ll 1$. In this case it follows from Eq. (21) that $b(T) < 0$, and one finds from Eq. (19),

$$\delta\;\cong\;\frac{b^2}{a^2} \tag{24}$$

Thus, δ is small but finite at $V \to \infty$, and μ remains smaller than m in the TL. The temperature dependence of chemical potential $\mu = \mu(T)$ for $V = 10^2$ fm^3 and 10^3 fm^3 at fixed pion number density ρ is shown in Fig. 2.

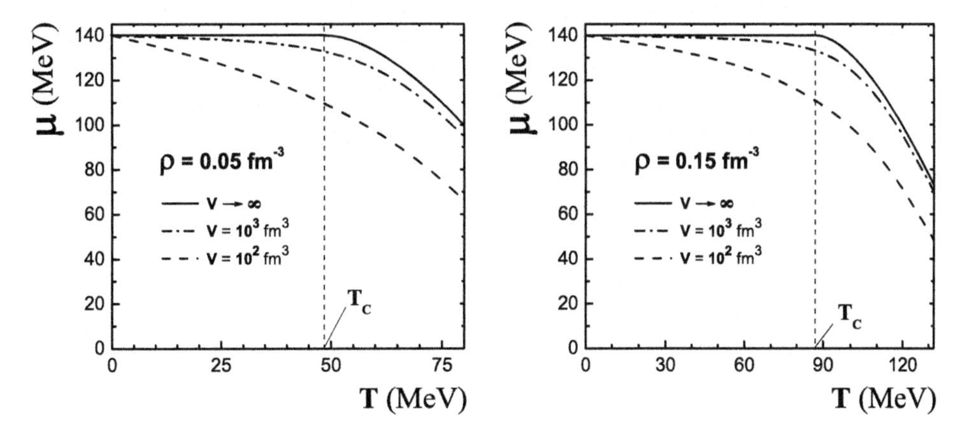

Fig. 2. The chemical potential μ as a function of temperature T at fixed particle number density ρ. The solid line presents the behavior in the TL $V \to \infty$. The dashed line corresponds to $V = 10^2$ fm^3, and dashed-dotted line to $V = 10^3$ fm^3. The vertical dotted line indicates the BEC temperature T_C. The left panel corresponds to $\rho = 0.05$ fm^{-3}, the right one to $\rho = 0.15$ fm^{-3}.

3.2. *Particle Number Fluctuations*

The variance of particle number fluctuations in the GCE at finite V is:

$$\begin{aligned} \langle \Delta N^2 \rangle \;&\equiv\; \langle (N \;-\; \langle N \rangle)^2 \rangle \;=\; \sum_{\mathbf{p},j} \langle n_{\mathbf{p},j} \rangle (1 + \langle n_{\mathbf{p},j} \rangle) \\ &\cong\; \frac{3}{\exp\left[(m-\mu)/T\right] \;-\; 1} + \frac{3}{\left\{\exp\left[(m-\mu)/T\right] \;-\; 1\right\}^2} \\ &+\; \frac{3V}{2\pi^2} \int_0^{\infty} p^2 dp \, \frac{\exp\left[\left(\sqrt{m^2+p^2}-\mu\right)/T\right]}{\left\{\exp\left[\left(\sqrt{m^2+p^2}-\mu\right)/T\right] \;-\; 1\right\}^2} \end{aligned} \tag{25}$$

where the first two terms in the r.h.s. of Eq. (25) correspond to particles of the lowest level $p = 0$, and the third term to particles with $p > 0$. We will use the scaled variance,

$$\omega \;=\; \frac{\langle \Delta N^2 \rangle}{\langle N \rangle} \tag{26}$$

as the measure of particle number fluctuations. The numerical results for the scaled variance are shown in Fig. 3. At $T > T_C$ the parameter δ goes to the finite limit

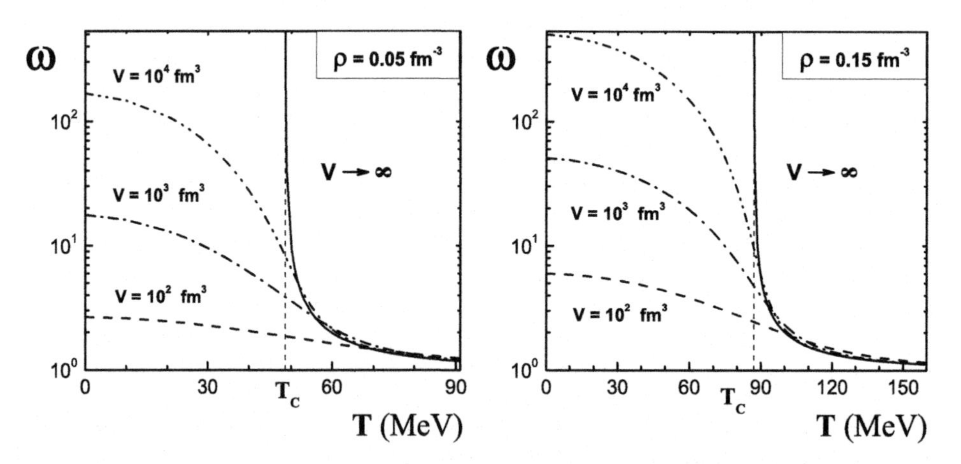

Fig. 3. The dashed lines show the GCE scaled variance (26) for the pion gas as a function of temperature T for $V = 10^4$ fm^3, 10^3 fm^3, 10^2 fm^3 (from top to bottom). The vertical dotted line indicates the BEC temperature T_C. The solid line shows the ω (26) in the TL $V \to \infty$. The left panel corresponds to $\rho = 0.05$ fm^{-3}, the right one to $\rho = 0.15$ fm^{-3} .

(24) at $V \to \infty$. This leads to the finite value of ω (26) in the TL. At $T \leq T_C$ one finds from Eq. (25) in the TL,

$$\langle \Delta N^2 \rangle \;\cong\; 3\,\delta^{-2} \;+\; \frac{3}{2}\; V\; a\; \delta^{-1/2} \tag{27}$$

This gives,

$$\omega \equiv \frac{\langle \Delta N^2 \rangle}{\langle N \rangle} \cong 3\,\rho^{-1}\,V^{-1}\,\delta^{-2} + \frac{3}{2}\,a\,\rho^{-1}\,\delta^{-1/2} \tag{28}$$

where $a = a(T)$ is defined in Eq. (20). The substitution of δ in Eq. (28) from (22), gives for $T < T_C$ and $V \to \infty$,

$$\omega \cong 3\,b^2\,\rho^{-1}\,V + \frac{3}{2}\,a\,b^{1/2}\,\rho^{-1}\,V^{1/2} \equiv \omega_C + \omega^* \tag{29}$$

The ω_C in the r.h.s. of Eq. (29) is proportional to V and corresponds to the particle number fluctuations in the BE condensate, i.e. at the $p = 0$ level, $\omega_C \cong \sum_j \langle (\Delta n_{p=0,j})^2 \rangle / \langle N \rangle$. The second term, ω^* is proportional to $V^{1/2}$. It comes from the fluctuation of particle numbers at $p > 0$ levels, $\omega^* = \sum_{\mathbf{p},j;p>0} \langle (\Delta n_{\mathbf{p},j})^2 \rangle / \langle N \rangle$. At $T \to 0$, one finds $a \to 0$ and $b \to \rho/3$. This gives the maximal value of the scaled variance, $\omega = \rho V/3 = \langle N \rangle /3$, for given ρ and V values.

The substitution of δ in Eq. (28) from (23), gives for $T = T_C$ and $V \to \infty$,

$$\omega \cong 3\,a^{4/3}\,\rho^{-1}\,V^{1/3} + \frac{3}{2}\,a^{4/3}\,\rho^{-1}\,V^{1/3} \equiv \omega_C + \omega^* \tag{30}$$

The Fig. 4 demonstrates the ratios ρ_C/ρ and ω_C/ω as the functions of T for $V = 10^2$, 10^3, 10^4 fm^3, and at $V \to \infty$. In the TL $V \to \infty$, one finds $\rho_C \to 0$ at $T \geq T_C$. The value of ρ_C starts to increase from zero at $T = T_C$ to ρ at $T \to 0$. Thus, ρ_C remains a continuous function of T in the TL. In contrast to this, both ω_C and ω^* have discontinuities at $T = T_C$. They both go to infinity in the TL $V \to \infty$. The ω_C/ω ratio equals to zero at $T > T_C$, 'jumps' from 0 to 2/3 at $T = T_C$, and further continuously approaches to 1 at $T \to 0$. At $T = T_C$ the contribution of $p = 0$ level to particle density, ρ_C, is negligible at $V \to \infty$, but the scaled variance ω_C from this level equals 2/3 of the total scaled variance ω and diverges as $V^{1/3}$. We conclude this section by stressing that the particle number fluctuations expressed by the scaled variance ω looks as a very promising quantity to search for the BEC in the pion gas.

4. BEC Fluctuation Signals in High Multiplicity Events

In the GCE, the scaled variances for different charge pion states, $j = +, -, 0$, are equal to each other and equal to the scaled variance ω for total number of pions,

$$\omega^j = 1 + \frac{\sum_{\mathbf{p},j} \langle n_{\mathbf{p},j} \rangle^2}{\sum_{\mathbf{p},j} \langle n_{\mathbf{p},j} \rangle} = \omega \tag{31}$$

There is a qualitative difference in the properties of the mean multiplicity and the scaled variance of multiplicity distribution in statistical models. In the case of the mean multiplicity results obtained with the GCE, canonical ensemble, and micro-canonical ensemble (MCE) approach each other in the TL. One refers here to the thermodynamical equivalence of the statistical ensembles. It was recently found[21–24] that corresponding results for the scaled variance are different in different

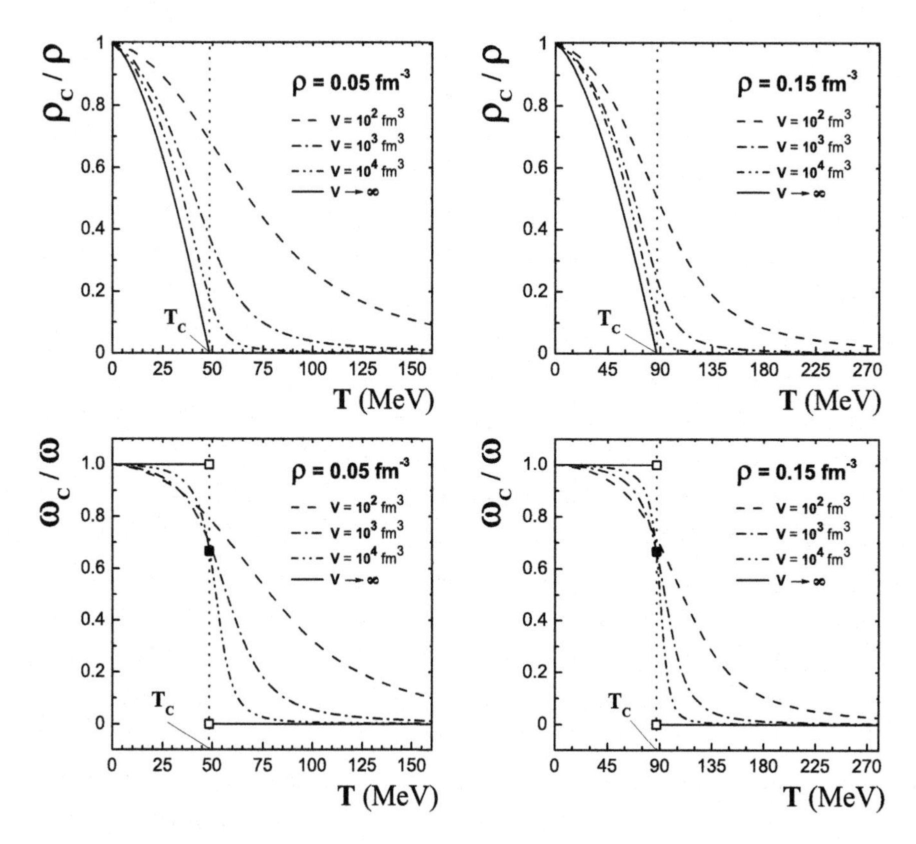

Fig. 4. The upper panel shows the ratio of condensate particle number density to the total particle number density, ρ_C/ρ, as functions of T for $V = 10^2$, 10^4, 10^4 fm^3, and in the TL $V \to \infty$. The lower panel shows the ratio of particle number fluctuations in condensate to the total particle number fluctuations, ω_C/ω, as functions of T for the same volumes. The vertical dotted line indicates the BEC temperature T_C. The left panel corresponds to $\rho = 0.05$ fm^{-3}, the right one to $\rho = 0.15$ fm^{-3} .

ensembles, and thus the scaled variance is sensitive to conservation laws obeyed by a statistical system. The differences are preserved in the thermodynamic limit. Therefore, the pion number densities are the same in different statistical ensembles, but this is not the case for the scaled variances of pion fluctuations. The pion number fluctuations in the system with fixed electric charge, $Q = 0$, total pion number, N, and total energy, E, should be treated in the MCE. The volume V is one more MCE parameter.

The MCE microscopic correlators equal to (see also Refs.[4,22]):

$$\begin{aligned}\langle \Delta n_{\mathbf{p},j} \Delta n_{\mathbf{k},i} \rangle_{mce} &= v^2_{\mathbf{p},j}\, \delta_{\mathbf{pk}} \delta_{ji} \\ &- v^2_{\mathbf{p},j}\, v^2_{\mathbf{k},i} \left[\frac{q_j q_i}{\Delta(q^2)} + \frac{\Delta(\epsilon^2) + \epsilon_{\mathbf{p}}\epsilon_{\mathbf{k}}\, \Delta(\pi^2) - (\epsilon_{\mathbf{p}} + \epsilon_{\mathbf{k}})\Delta(\pi\epsilon)}{\Delta(\pi^2)\Delta(\epsilon^2) - (\Delta(\pi\epsilon))^2} \right]\end{aligned} \tag{32}$$

where $q_+ = 1$, $q_- = -1$, $q_0 = 0$, $\Delta(q^2) = \sum_{\mathbf{p},j} q_j^2 \upsilon_{\mathbf{p},j}^2$, $\Delta(\pi^2) = \sum_{\mathbf{p},j} \upsilon_{\mathbf{p},j}^2$, $\Delta(\epsilon^2) = \sum_{\mathbf{p},j} \epsilon_{\mathbf{p}}^2 \upsilon_{\mathbf{p},j}^2$, $\Delta(\pi\epsilon) = \sum_{\mathbf{p},j} \epsilon_{\mathbf{p}} \upsilon_{\mathbf{p},j}^2$. Note that the first term in the r.h.s. of Eq. (32) corresponds to the GCE (2). From Eq. (32) one notices that the MCE fluctuations of each mode $\mathbf{p}$ are reduced, and the (anti)correlations between different modes $\mathbf{p} \neq \mathbf{k}$ and between different charge states appear. This results in a suppression of scaled variance ω_{mce} in a comparison with the corresponding one ω in the GCE. Note that the MCE microscopic correlators (32), although being different from that in the GCE, are expressed with the quantities calculated in the GCE. The straightforward calculations lead to the following MCE scaled variance for π^0-mesons[4] :

$$\omega_{mce}^{0} = \frac{\sum_{\mathbf{p},\mathbf{k}} \langle \Delta n_{\mathbf{p},0} \, \Delta n_{\mathbf{k},0} \rangle_{mce}}{\sum_{\mathbf{p}} \langle n_{\mathbf{p},0} \rangle} \cong \frac{2}{3} \, \omega \tag{33}$$

Due to conditions, $N_+ \equiv N_-$ and $N_+ + N_- + N_0 \equiv N$, it follows, $\omega_{mce}^{\pm} = \omega_{mce}^0/4 = \omega/6$ and $\omega_{mce}^{ch} = \omega_{mce}^0/2 = \omega/3$, where $N_{ch} \equiv N_+ + N_-$.

The pion number fluctuations can be studied in high energy particle and/or nuclei collisions. To search for the BEC fluctuation signals one needs the event-by-event identifications of both charge and neutral pions. Unfortunately, in most event-by-event studies, only charge pions are detected. In this case the global conservation laws would lead to the strong suppression of the particle number fluctuations, see also Refs.[4,5] , and no anomalous BEC fluctuations would be seen.

As an example we consider the high π-multiplicity events in $p+p$ collisions at the beam energy of 70 GeV (see Ref.[25]). In the reaction $p+p \rightarrow p+p+N$ with small final proton momenta in the c.m.s., the total c.m. energy of created pions is $E \cong \sqrt{s} - 2m_p \cong 9.7$ GeV. The estimates[26] reveal a possibility to accumulate the samples of events with fixed $N = 30 \div 50$ and have the full pion identification. Note that for this reaction the kinematic limit is $N^{max} = E/m_\pi \cong 69$. To define the MCE pion system one needs to assume the value of V, in addition to given fixed values of $Q = 0$, $E \cong 9.7$ GeV, and N. The T and μ parameters of the GCE can be then estimated from the following equations,

$$E = V \, \varepsilon(T,\mu;V) \, , \quad N = V \, \rho(T,\mu;V) \tag{34}$$

In calculating the ε and ρ in Eq. (34) we take into account the finite volume effects according to Eqs. (16-17) as it is discussed in Sec. II. Several 'trajectories' with fixed energy density are shown in Fig. 5 starting from the line $\mu = 0$ in the pion gas in the $\rho - T$ phase diagram. The MCE scaled variance of π^0 number fluctuations, ω_{mce}^0, increases with increasing of N. The maximal value it reaches at $T \rightarrow 0$,

$$\omega_{mce}^{0 \; max} \cong \frac{2}{3} \left(1 \, + \, \langle N_0 \rangle^{max}\right) = \frac{2}{3}\left(1 \, + \, \frac{N^{max}}{3}\right) \cong 16 \tag{35}$$

In Fig. 6, ω_{mce}^0 is shown as the function of N. Different possibilities of fixed energy densities and fixed particle number densities are considered. One way or

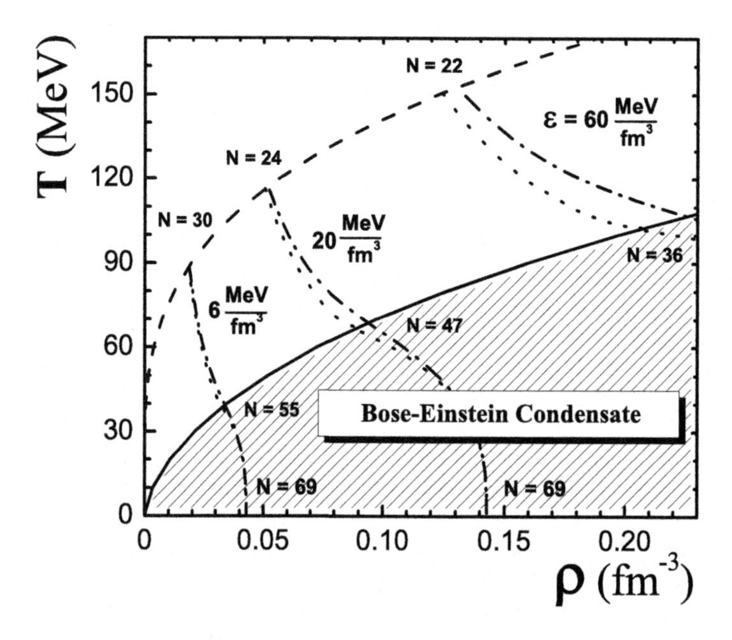

Fig. 5. The phase diagram of the ideal pion gas with zero net electric charge. The dashed line corresponds to $\rho = \rho^*(T, \mu = 0)$ and the solid line to the BEC $T = T_C$ (4), both calculated in the TL $V \rightarrow \infty$. The dashed-dotted lines present the trajectories in the $\rho - T$ plane with fixed energy densities, $\varepsilon = 6,\ 20,\ 60$ MeV/fm^3, calculated for the finite pion system with total energy $E = 9.7$ GeV according to Eq. (17). The dotted lines show the same trajectories calculated in the TL $V \rightarrow \infty$. The total numbers of pions N marked along the dashed-dotted lines correspond to 3 points: $\mu = 0,\ T = T_C$, and $T = 0$ for $E = 9.7$ GeV.

another, an increase of N leads to a strong increase of the fluctuations of N_0 and N_{ch} numbers due to the BEC effects.

The large fluctuations of $N_0/N_{ch} = f$ ratio were also suggested (see, e.g., Ref.[27]) as a possible signal for the disoriented chiral condensate (DCC). The DCC leads to the distribution of f in the form, $dW(f)/df = 1/(2\sqrt{f})$. The thermal Bose gas corresponds to the f-distribution centered at $f = 1/2$. Therefore, f-distributions from BEC and DCC are very different, and this gives a possibility to distinguish between these two phenomena.

5. Summary

The idea for searching the pion BEC as an anomalous increase of the pion number fluctuations was suggested in our previous papers[4,5] . The fluctuation signals of the BEC have been discussed in Refs.[4,5] in the thermodynamic limit. At $V \rightarrow \infty$, it follows, $\omega = \infty$ at $T \leq T_C$. This is evidently not the case for the finite systems. At finite V the scaled variance ω of the pion number fluctuation is finite for all possible combinations of the statistical system parameters. The ω demonstrates different dependence on the system volume V in different parts of the $\rho - T$ phase diagram.

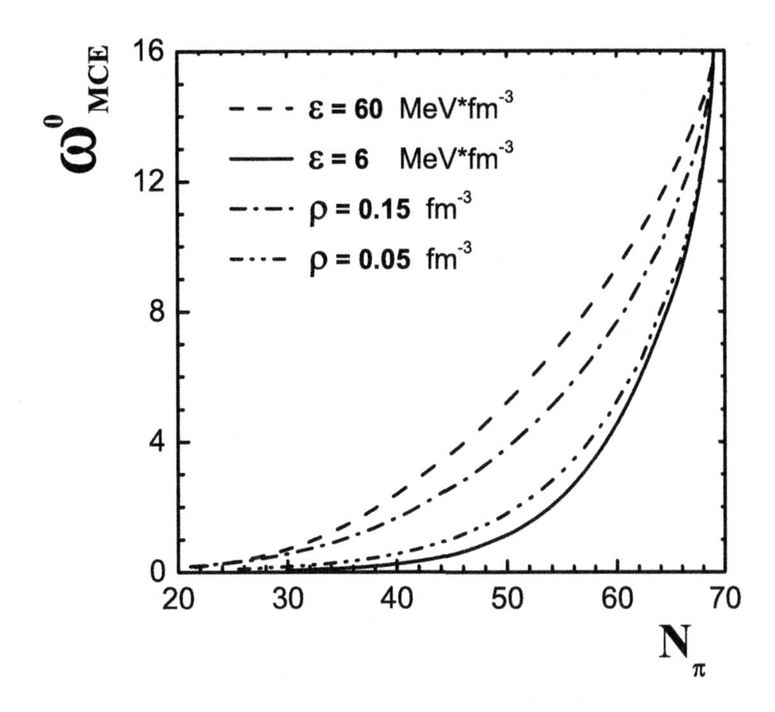

Fig. 6. The scaled variance of neutral pions in the MCE is presented as the function of the total number of pions N. Three solid lines correspond to different energy densities, $\varepsilon = 6,\ 60\ \mathrm{MeV/fm^3}$ (from bottom to top), calculated according to Eq. (17). Two dashed-dotted lines correspond to different particle number densities, $\rho = 0.05,\ 0.15\ \mathrm{fm^{-3}}$ (from bottom to top), calculated according to Eq. (16). The scaled variance ω^0_{mce} is given by Eq. (33), with ω (26) and $\langle \Delta N^2 \rangle$ (25). The total energy of the pion system is assumed to be fixed, $E = 9.7$ GeV.

In the TL $V \to \infty$, it follows that ω converges to a finite value at $T > T_C$. It increases as $\omega \propto V^{1/3}$ at the BEC line $T = T_C$, and it is proportional to the system volume, $\omega \propto V$, at $T < T_C$. The statistical model description gives no answer on the value of V for given E and N. The system volume remains a free model parameter. Thus, the statistical model does not suggest an exact quantitative predictions for the N-dependence of ω^0_{mce} and $\omega^{\pm}_{mce}$ in the sample of high energy collision events. However, the qualitative prediction looks rather clear: with increasing of N the pion system approaches the conditions of the BEC. One observes an anomalous increase of the scaled variances of neutral and charged pion number fluctuations. The size of this increase is restricted by the finite size of the pion system. In turn, a size of the created pion system (maximal possible values of N and V) should increase with the collision energy.

Acknowledgments

We would like to thank A.I. Bugrij, M. Gaździcki, W. Greiner, V.P. Gusynin, M. Hauer, B.I. Lev, I.N. Mishustin, St. Mrówczyński, M. Stephanov, and E. Shuryak for discussions. We are also grateful to E.S. Kokoulina and V.A. Nikitin for the information concerning to their experimental project[25] . The work was supported in part by the Program of Fundamental Researches of the Department of Physics and Astronomy of NAS Ukraine. V.V. Begun would like also to thank for the support of The International Association for the Promotion of Cooperation with Scientists from the New Independent states of the Former Soviet Union (INTAS), Ref. Nr. 06-1000014-6454.

References

1. S. N. Bose, *Z. Phys.* **26**, 178 (1924); A. Einstein, *Sitz. Ber. Preuss. Akad. Wiss. (Berlin)* **1**, 3 (1925).
2. M. H. Anderson, et al., *Science* **269**, 198 (1995); K.B. Davis, et al., *Phys. Rev. Lett.* **75**, 3969 (1995).
3. J. Zimanyi, G. Fai, and B. Jakobsson, *Phys. Rev. Lett.* **43**, 1705 (1979); I. N. Mishustin, et al., *Phys. Lett. B* **276**, 403 (1992); C. Greiner, C. Gong, and B. Müller, *ibid* **316**, 226 (1993); S. Pratt, *Phys. Lett. B* **301**, 159 (1993); T. Csörgö and J. Zimanyi, *Phys. Rev. Lett.* **80**, 916 (1998); A. Bialas and K. Zalewski, *Phys.Rev. D* **59**, 097502 (1999); Yu. M. Sinyukov, S. V. Akkelin, and R. Lednicky, nucl-th/9909015; R. Lednicky, et al., *Phys. Rev. C* **61** 034901 (2000).
4. V. V. Begun and M. I. Gorenstein, *Phys. Lett. B* **653**, 190 (2007).
5. V. V. Begun and M. I. Gorenstein, arXiv:0709.1434 [hep-ph].
6. V. V. Begun and M. I. Gorenstein, arXiv:0802.3349 [hep-ph].
7. L. D. Landau and E. M. Lifshitz. *Statistical Physics (Course of Theoretical Physics, Volume 5).* Pergamon Press Ltd. 1980.
8. H. E. Haber and H. A. Weldon, *Phys. Rev. Lett.* **46**, 1497 (1981); *Phys. Rev. D* **25**, 502 (1982).
9. J. I. Kapusta, *Finite-Temperature Field Theory*, Cambridge, 1989.
10. L. Salasnich, *Nuovo Cim. B* **117** 637 (2002).
11. D. T. Son and M. A. Stephanov, *Phys. Rev. Lett.* **86** 592 (2001); *Phys. Atom. Nucl.* **64** 834 (2001); [*Yad. Fiz.* **64** 899 (2001)]; K. Splittorff, D. T. Son and M. A. Stephanov, *Phys. Rev. D* **64** 016003 (2001).
12. V. V. Begun and M. I. Gorenstein, *Phys. Rev. C* **73**, 054904 (2006).
13. J. Madsen, *Phys. Rev. Lett.* **69**, 571 (1992); J. Madsen, *Phys. Rev. D* **64**, 027301 (2001).
14. D. Boyanovsky, H. J. de Vega, and N. G. Sanchez, arXiv:0710.5180[astro-ph].
15. M. Abramowitz and I. E. Stegun, *Handbook of Mathematical Functions* (Dover, New York, 1964).
16. V. V. Tolmachjev, *Theory of Bose Gas*, Moscow University, 1969 (in Russian).
17. Yu. B. Rumer and M. Sh. Ryvkin, *Thermodynamics, Statistical Physics, and Kinetics*, Nauka, 1972 (in Russian).
18. W. Greiner, L. Neise, and H Stöcker, *Thermodynamics and Statistical Mechanics*, 1995 Springer-Verlag New York, Inc.
19. A. P. Prudnikov, Yu. A. Brychkov, and O. I. Marichev, *Integrals and Series*, (Moscow, Nauka, 1986).
20. K. B. Tolpygo, *Thermodynamics and Statistical Physics*, Kiev University, 1966 (in Russian).
21. V. V. Begun, et al., *Phys. Rev. C* **70**, 034901 (2004); *ibid* **71**, 054904 (2005); *ibid* **72**, 014902 (2005); *J. Phys. G* **32**, 935 (2006); A. Keränen, et al., *J. Phys. G* **31**, S1095

(2005); F. Becattini, et al., *ibid* **72**, 064904 (2005); J. Cleymans, K. Redlich, and L. Turko, *ibid* **71**, 047902 (2005); *J. Phys. G* **31**, 1421 (2005).

22. V. V. Begun, et al., *Phys. Rev. C* **74**, 044903 (2006); *ibid* **76**, 024902 (2007).
23. M. Hauer, V. V. Begun, and M. I. Gorenstein, arXiv:0706.3290 [nucl-th].
24. M. I. Gorenstein, arXiv:0709.1428 [nucl-th]; V. V. Begun, arXiv:0711.2912 [nucl-th].
25. P. F. Ermolov, et al., *Phys. At. Nucl.* **67**, 108 (2004); V. V. Avdeichikov, et al., JINR-P1-2004-190, 45 pp (2005).
26. V. A. Nikitin, private communication.
27. J. P. Blaizot and A. Krzywicki, *Phys. Rev. D* **46**, 246 (1992); *Acta Phys. Pol. B* **27**, 1687 (1996); K. Rajagopal and F. Wilczek, *Nucl. Phys. B* **399**, 395 (1993); J. Bjorken, *Acta Phys. Pol. B* **28**, 2773 (1997).

OBSERVATION OF A RESONANCE-LIKE STRUCTURE IN THE INVARIANT MASS SPECTRUM OF TWO PHOTONS FROM pC- AND dC-INTERACTIONS

K. U. ABRAAMYAN[a,b], M. I. BAZNAT[c], A. V. FRIESEN[a], K. K. GUDIMA[c],
M. A. KOZHIN[a], S. A. LEBEDEV[d,e], M. A. NAZARENKO[a,f], S. A. NIKITIN[a],
G. A. OSOSKOV[d], S. G. REZNIKOV[a], A. N. SISSAKIAN[g], A. S. SORIN[g], and
V. D. TONEEV[g]

a) VBLHE JINR, 141980 Dubna, Moscow region, Russia
b) Yerevan State University, Yerevan, Armeniya
c) Institute of Applied Physics, Kishinev, Moldova
d) LIT JINR, 141980 Dubna, Moscow region, Russia
e) Gesellschaft für Schwerionenforschung, Darmstadt, Germany
f) Moscow State Institute of Radioengineering, Electronics and Automation, 119454 Moscow, Russia
g) BLTP JINR, 141980 Dubna, Moscow region, Russia

Along with π^0 and η mesons, a resonance structure in the invariant mass spectrum of two γ-quanta at $M = 360 \pm 7 \pm 9$ MeV is first observed in the reaction $dC \to \gamma + \gamma + X$ at momentum 2.75 GeV/c per nucleon. Preliminary estimates of its width and cross section are $\Gamma = 49.2 \pm 18.6$ MeV and $\sigma_{\gamma\gamma} \sim 98$ μb. The collected statistics is 2339 ± 340 events of $1.5 \cdot 10^6$ triggered interactions of a total number of $\sim 10^{12}$ dC-interactions. This structure is not observed in pC collisions at the beam momentum 5.5 GeV/c. Possible mechanisms of this ABC-like effect are discussed.

1. Introduction

Dynamics of the near-threshold pion production and $\pi\pi$ interaction is of imperishable interest. Among the oldest and still puzzling problems there is the so-called "ABC effect". Almost fifty years ago Abashian, Booth, and Crowe[1] have first observed anomaly in the production of pion pairs in the reaction $dp \to ^3$He$+2\pi \equiv ^3$He$+X^0$. This anomaly or ABS effect stands for an unexpected enhancement in the spectrum of the invariant $\pi\pi$ mass at masses of about 40 MeV higher than $2m_\pi$. Following experiments $dp \to ^3$He$+X$[2] , $pn \to d + 2\pi$[3-5] , $dd \to ^4$He$+X$[6,7] , $np \to d + 2\pi$ with neutron beams[8,9] and even $np \to d + \eta$[10] independently confirmed this finding. Such an anomaly was also observed in the photoproduction of pion pairs, $\gamma p \to p + X^0$[11,12] . It was revealed that the ABC effect is of isoscalar nature since a similar effect was not observed in the $pd \to H^3 + X^+$ reaction. The masses of the peak position and its widths vary for different bombarding energies and observation angles. Initially the low-mass enhancement has been interpreted as caused by an unusually strong s-wave $\pi\pi$ interaction or as evidence for the σ meson existence[1] . It is usually accepted now that this enhancement is not an intrinsic two-pion property since there is no resonance structure in the $\pi\pi$ scattering amplitude in this energy range. So any interpretation of the ABC as a real resonance is very much in doubt (for example, see discussion in[8]). It is generally believed that system like that has to be associated with two nucleons when two pions (both must be present) are rescattered on them or both nucleons participate in elementary $pp \to \pi + X$ reactions (predominantly via Δ formation). Actually, the origin of the ABC effect must be looked for in the formation of light nuclei at intermediate energies (for a review see Ref.[13]).

All experiments conducted on this issue with the exception of low-statistics bubble-chamber measurements[5,9] were inclusive measurements carried out preferentially with a single-armed magnetic spectrograph for detection of the fused nuclei, which allows one to find the two-pion invariant mass through the missing mass. Very recently, exclusive measurements of the reactions $pd \to p + d + \pi^0 + \pi^0$ and $pd \to^3$He$+\pi + \pi$ have been carried out with complete kinematics in the energy range of the ABC effect at CELSIUS using the 4π WASA detector setup[14,15] . The importance of the strongly attractive $\Delta\Delta$ channel was noted. Surprisingly, the basic $pp \to pp\pi^0\pi^0$ reaction in the $\Delta\Delta$ region also shows an ABC-like low-$\pi\pi$ mass enhancement, which deserves special attention. It confirms the earlier result in[16] where the analyzing powers and cross sections for the ABC enhancement production were measured for the reaction $\vec{p}p \to p + p + X^0$ in the missing mass range $m_\pi^0 < M_{X^0} < m_\eta$. However, the big difference was observed in the width of the resonant cross section and it was concluded that the observed width in the isoscalar channel is not obviously just a simple result of the binding between the two Δ states. It rather signals more complicated configurations in the wave function of the intermediate state, as would be expected for a nontrivial dibaryon state[15] .

This work aims to study whether this low-$\pi\pi$ mass anomaly can survive in heavier systems in the $\gamma\gamma$ channel. The paper is organized as follows. After a brief description of the experiment and experimental setup, the structure of measured invariant mass spectra of two photon pairs is analyzed in Sect.2. As a cross-check, a similar analysis is carried out in Sect.3 but within the wavelet method. To elucidate the nature of the peak at $M_{\gamma\gamma} \approx (2-3)m_\pi$, different mechanisms of the observed $\gamma\gamma$ pair enhancement are discussed in Sect.4. In Sect.5, experimental estimates for production cross sections and widths of η mesons and hypothetical R resonance are given. The main inferences of the paper are presented in the Conclusions.

2. Experiment

2.1. *General layout*

The data acquisition of production of neutral mesons and γ-quanta in pC and dC interactions has been carried out with internal beams of the JINR Nuclotron[17,18] . Experiments were conducted with internal proton beams at momentum 5.5 GeV/c

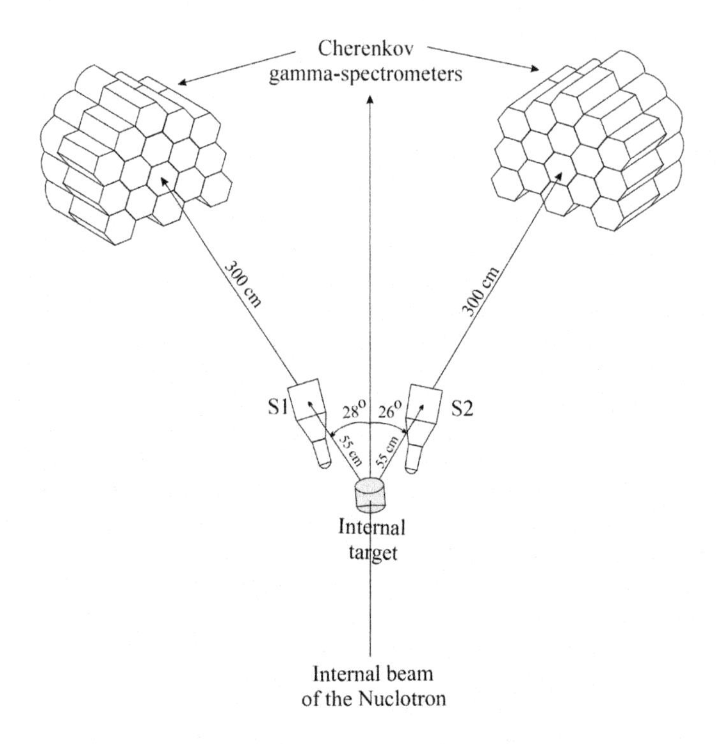

Fig. 1. The schematic drawing of the experimental PHOTON-2 setup. The S_1 and S_2 are scintillation counters.

incident on a carbon target and with ^2H, ^4He beams and internal C-, Al-, Cu-, W-, Au-targets at momenta from 1.7 to 3.8 GeV/c per nucleon. For the first analysis the data with the maximal statistics, pC- and dC-interactions, were selected. The first preliminary results on $dC(2\ GeV)$ collisions were represented at the meeting[18] .

The presented data concern reactions induced by deuterons with a momentum 2.75 GeV/c per nucleon and of protons with 5.5 GeV/c. Typical deuteron and proton fluxes were about 10^9 and $2 \cdot 10^8$ per pulse, respectively. The electromagnetic lead glass calorimeter PHOTON-2 was used to measure both the energies and emission angles of photons. The results obtained in earlier experiments with this setup are published in[19] . The experimental instrumentation is schematically represented in Fig.1.

The PHOTON-2 setup includes 32 γ-spectrometers of lead glass and scintillation counters S_1 and S_2 of $2 \times 15 \times 15\ cm^3$ used in front of the lead glass for the charged particle detection[19–21] .

The center of the front surfaces of the lead glass hodoscopes is located at 300 cm from the target and at angles 25.6^0 and 28.5^0 with respect to the beam direction. This gives a solid angle of 0.094 sr (0.047 sr for each arm). Some details of the

Table 1. The basic parameters of the lead glass hodoscope.

Number of lead glasses	32 TF-1, total wight 1422 kg
Module cross section	r=9 cm of insert circumference
Module length	35 cm, 14 R.L.
Spatial resolution	3.2 cm
Angular resolution	0.6^0
Energy resolution	$(3.9/\sqrt{E}+0.4)\%$, E[GeV]
Gain stability	(1-2)%
Dynamic range	50 MeV - 6 GeV
Minimum ionizing signal	382 $\pm$4 MeV of the photon equivalent
Total area	0.848 m^2

construction and performance of the lead glass hodoscope are given in Table.2.1. The internal target is a rotating (in a vertical plane) wire with the diameter of 8 micron. Duration of a cycle is 1 second.

The modules of the γ-spectrometer are assembled into two arms of 16 units. These modules in each arm are divided into two groups of 8 units. The output signals in each group are summed up linearly and after discrimination by amplitude are used in fast triggering. In this experiment the discriminator thresholds were at the level of 0.4 GeV. Triggering takes place when there is a coincidence of signals from two or more groups from different arms. The mean rate of triggering was about 330 and 800 events per spill in dC and pC reactions, respectively. Totally about 1.52×10^6 and $1.06 \cdot 10^6$ triggers were recorded during these experiments.

2.2. *Event selection*

Photons in the detector are recognized as isolated and confined clusters (an area of adjacent modules with a signal above the threshold) in the electromagnetic calorimeter. The photon energy is calculated from the energy of the cluster. Assuming that the photon originates from the target, its direction is determined from the geometrical positions of constituent modules weighted with the corresponding energy deposit in activated modules.

The invariant mass distributions of photon pairs (from different arms of the spectrometer) are shown in Fig.2. The dominant part of distributions (two upper panels) comes from the $\pi^0 \to \gamma\gamma$ decay which suppresses the expected contribution of the $\eta \to \gamma\gamma$ decay. As is seen, the high-energy cut of photons, $E_\gamma > 100$ MeV, allows one to improve the situation with signal/background ratio.

To see a possible structure of the invariant mass spectra, a background should be subtracted. The so-called event mixing method was used to estimate the combinatorial background: combinations of γ-quanta were sampled randomly from different events. For the general sampling and combinatorial background one, the same following selection criteria were used:

(1) the number of γ-quanta in an event, $N_\gamma = 2$;
(2) the energies of γ-quanta, $E_\gamma \geq 100$ MeV;

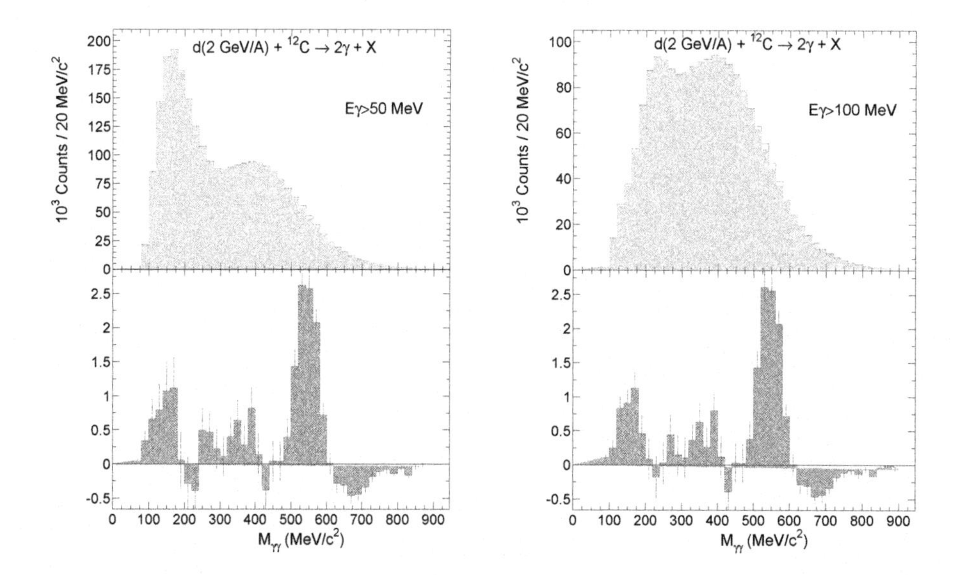

Fig. 2. Invariant mass distribution of $\gamma\gamma$ pairs from the reaction $dC \to \gamma + \gamma + X$ at 2.75 GeV/c per nucleon for two values of the cut energy of photons. The top shaded histograms show the background contribution. The bottom histograms are invariant spectra after the background subtraction.

(3) the summed energy in real and random events ≤ 1.5 GeV.

A background like this was subtracted from the invariant mass distributions (see bottom panels in Fig.2. A clear peak from the η decay and a remnant of the suppressed π^0 resonance are clearly observable. Note that between them there is some additional structure which will be clarified below.

Systematic errors may be due to uncertainty in measurements of γ energies and inaccuracy in estimates of the combinatorial background. The method of energy reconstruction of events is described in detail in Refs.[19,20] . One of the criteria of accuracy of energy reconstruction is the conformity of the peak positions corresponding to the known particle mass values. As is seen in Fig.2, the position of peaks corresponding to η- and π^0-mesons is in reasonable agreement with the table values of their masses. A more precise determination of the position of peaks requires minimization of systematic errors in describing the background which arise, in particular, due to the violation of the energy-momentum conservation laws in selecting γ-quanta by random sampling from different events (see also below).

The selection criteria can be made harder by imposing the trigger condition described above: the sampling is triggered if two or more groups of modules with the total energy > 400 MeV are available. The result of this triggering is shown in Fig.3. Under this additional condition the π^0 peak is practically absent. Therefore, π^0-mesons were mainly registered in events with $N_\gamma > 2$ (a minimal opening angle

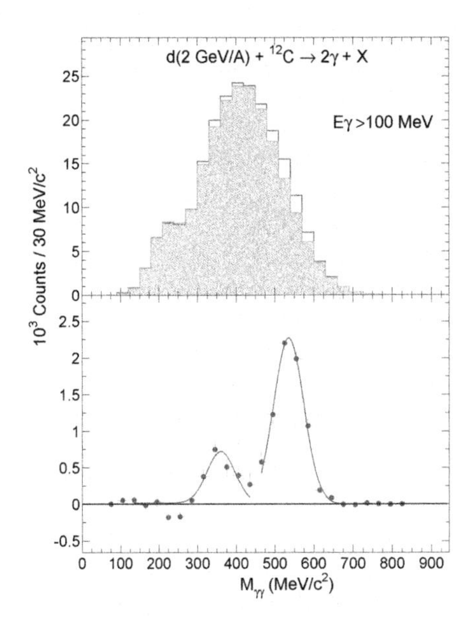

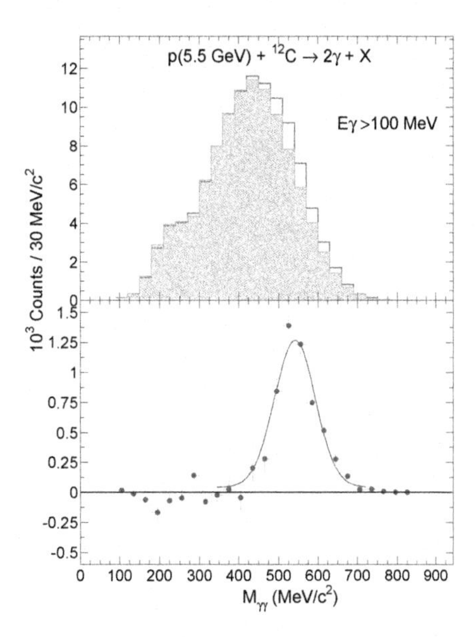

Fig. 3. Invariant mass distributions of $\gamma\gamma$ pairs satisfying criteria 1) − 3) and the trigger condition without (upper panel) and with (bottom panel) the background subtraction. The left and right figures are obtained for the reaction $dC \to \gamma + \gamma + X$ at 2.75 GeV/c per nucleon and pC collision at 5.5 GeV/s, respectively. The curves are the Gaussian approximation of experimental points (see the text).

Table 2. Fit parameters of the Gaussian distribution.

	$dC, M_{\gamma\gamma} \leq 435$	$dC, M_{\gamma\gamma} \geq 465$	$pC, M_{\gamma\gamma} \geq 465$ MeV
y_0	0.50± 0.77	-0.004± 0.046	1.44± 1.86
N_0	2116± 381	7329± 295	5283± 560
w_{measur}	36.10± 6.9	38.5± 1.7	51.6± 4.1
M_0	360.0± 7.0	535.7 ± 1.9	541.5 ± 2.5
$\chi^2\ nfr$	2.15	0.82	2.30

of the γ pair registered by the setup equals 42°). In contrast, the η is seen very distinctly with the width defined by the experimental resolution in the mass. In addition, in this reaction $dC \to \gamma + \gamma + X$ a pronounced peak is observed in the interval 300-420 MeV of the invariant mass two-photon spectrum. However, under similar conditions only η is seen in the pC collisions. The observed peaks were approximated independently by the Gaussian:

$$\frac{dN}{dM_{\gamma\gamma}} = y_0 + \frac{N_0}{w_{measur}\sqrt{2\pi}} \exp\left(-\frac{(M_{\gamma\gamma} - M_0)^2}{2w_{measur}^2}\right) \quad (1/MeV). \tag{1}$$

The additional shift-parameter y_0 is introduced in eq.(1). The values of the obtained fitting parameters are given in Table.2.2.

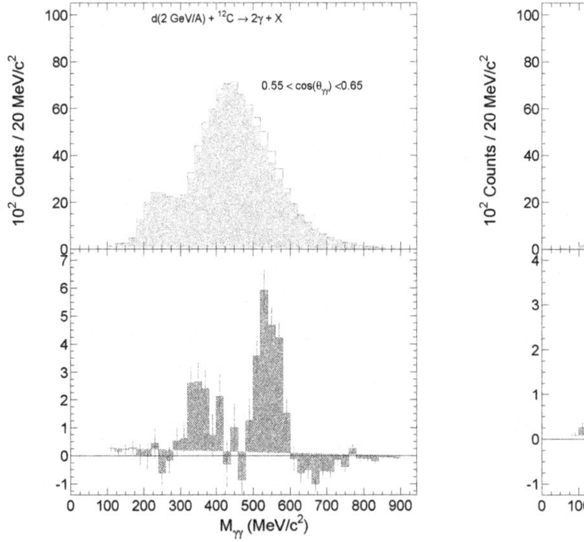

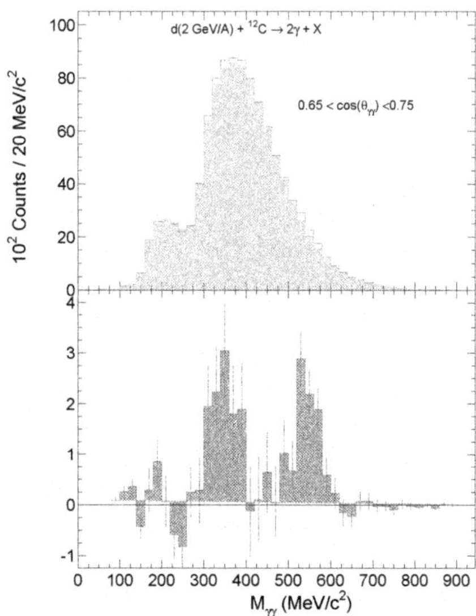

Fig. 4. The invariant mass distributions of two photons for the opening angles $0.55 < \cos\Theta_{\gamma\gamma} < 0.65$ (left) and $0.65 < \cos\Theta_{\gamma\gamma} < 0.75$ (right). Selection conditions are the same as in the right panel of Fig.2.

The signals/background ratios for the invariant mass intervals of 300-420 MeV and 480-600 MeV (the vicinity of the η-meson mass) are $2.5 \cdot 10^{-2}$ and $1.4 \cdot 10^{-1}$, respectively. For comparison, analogous values without the background suppression (without the selection criteria 1) and 2)) are $(4.0 \pm 1.4) \cdot 10^{-3}$ and $3.2 \cdot 10^{-2}$.

Thus, as follows from Tabl.2.2, the position and width of the peak corresponding to η-meson are in good agreement with values from the PDG table (systematic errors do not exceed 1.5%) and the spectrometer mass resolution. The total number of events registered in the η-meson region 450-660 MeV after background subtraction is 7336 ± 284.

To elucidate the nature of the detected enhancement, we investigate the dependence of its position and width on the opening angle of two photons and on the their energy selection level. The results are demonstrated in Figs. 4 and 5.

As is seen, the result of the changing of observation conditions is quite robust: the position and width of the observed peak remain almost unchanged in different intervals of both energies and opening angles of γ-quanta. The mean peak position in the invariant mass distribution varies under different conditions in the range $348 \div 365$ MeV. The total number of events registered in the region 270-450 MeV (a summed number of pairs in the histogram in Fig.3) after the background subtraction is 2339 ± 340.

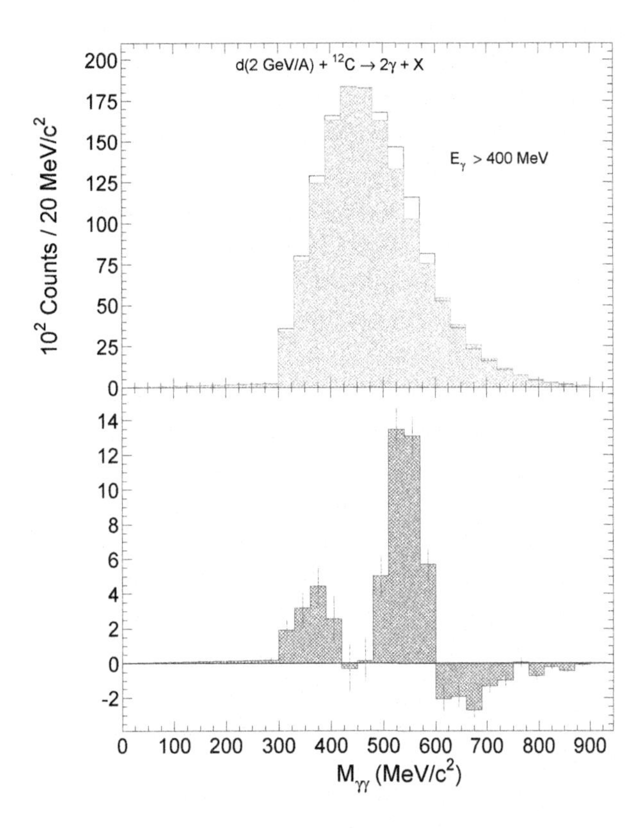

Fig. 5. The invariant mass spectra of $\gamma\gamma$ pairs for the energy selection $E_\gamma > 400$ MeV. Selection conditions are the same as in Fig.2.

3. Wavelet analysis

Here we shall try to identify essential structures in the measured $M_{\gamma\gamma}$ spectra without the background subtraction. The wavelet analysis allows one to do that. One dimensional wavelet transform (WT) of a signal f(x) has a biparametric form. This allows WT to overcome the main shortcomings of the Fourier transform such as nonlocality and infinite support, necessity of a broad band of frequencies to decompose even a short signal. The wavelet transformation changes the decomposition basis into functions which are compact into time/space and frequency domain, so WT with the wavelet function ψ of the function $f(x)$ is defined by their convolution as

$$W_\psi(a,b)f = \frac{1}{\sqrt{C_\psi}} \int_{-\infty}^{\infty} \frac{1}{\sqrt{|a|}} \psi\left(\frac{b-x}{a}\right) f(x)\, dx \tag{2}$$

with the normalization constant

$$C_\psi = 2\pi \int\limits_{-\infty}^{\infty} \frac{|\tilde{\psi}(\omega)|^2}{|\omega|}\, d\omega \; < \; \infty,$$

where $\tilde{\psi}(\omega)$ is the Fourier transform of the wavelet $\psi(x)$. The parameters a and b define the scale and shift of the WT. In this respect the wavelet function $\psi(t)$ is a sort of "window function" with a non-constant window's width: high-frequency events are narrow (due to the factor $1/a$), while low-frequency wavelets are broader. The inverse transform is given by the formula

$$f(t) = C_\psi^{-1} \int\int \psi\left(\frac{t-b}{a}\right) W_\psi(a,b)\, \frac{da\; db}{a^2} \,. \tag{3}$$

The condition $C_\psi < \infty$ is that for the wavelet ψ to exist. It holds, in particular, when the first $(n-1)$ momenta are equal to zero

$$\int\limits_{-\infty}^{\infty} |x|^m \psi(x) dx = 0, \quad 0 \leq m < n. \tag{4}$$

Due to freedom in the choice of the wavelet function ψ, many different wavelets were invented[22,24] , but we consider here only the so-called *continuous Wavelets with Vanishing Momenta* (WVM) (see Appendix). The VMW family is called so because condition (4) always holds for it. One of WVM families is named *Gaussian wavelets* (GW) because they are normalized derivatives of the well-known Gauss function

$$g(x; A, x_0) = A \exp\left(-\frac{(x-x_0)^2}{2\sigma^2}\right). \tag{5}$$

Such an advantage of wavelets as their ability to separate signal components with different frequencies and positions has attracted many physicists to use both discrete and continuous wavelets[27,28] . Usually, the conventional filtering approach is applied: a signal transformed by a wavelet underdoes an appropriate thresholding and then is restored by the inverse transform. The image of the wavelet spectrum is used to obtain rough parameter estimations of wanted peaks of invariant mass spectra, as in[29] , where the G_2 Mexican hat wavelet was used. In our paper, we apply the family of GW to look for peaks in question having a Gaussian shape (5). That makes it possible to use very simple analytical expressions (A.5) of the Gaussian wavelet transform for Gaussian peaks, which give us a remarkable advantage to calculate the peak parameters directly in the wavelet domain instead of time/space domain *without using the inversion.* Moreover, in real cases, when our signal shape is close to Gaussian and is considerably contaminated by additive noise and, in addition, distorted by binning to be input to computer, one can also use the remarkable robustness of Gaussian wavelet filtering, as proved in[24] .

The main idea of our approach is to transform the signal $f(x)$ to the space of the corresponding wavelet and look there for a local biparametric area surrounding the wavelet image of our peak in question, drawing down all other details of the signal image, concerning noise, binning effects and background pedestal. Thus, after the inverse transform of this local area one can expect to obtain the initial peak, cleaned from all those contaminations. Assuming that the peak in question has a Gaussian shape, we transform the given invariant mass spectrum into a chosen wavelet (say, G_2 or higher, up to G_8) domain where an easy analytical formula of the corresponding wavelet surface is valid for such Gaussian-like signals. Therefore, one can fit this ideal surface to that obtained for the real spectrum. To initiate this 2D nonlinear fit, one needs a starting point in the (a, b) domain. It is found in the following way. It looks obvious that in the shift dimension b the maximum of Gaussian (5) corresponds to the $G_n(a, b)$ value $b = x_0$. Then considering these expressions as functions of the scale parameter a, one can find their maxima as the solution of equations obtained by differentiating and zeroing each of them. In this way one would obtain just the 2D maximum of the corresponding wavelet.

Let us demonstrate this by $G_2(a, b)$ wavelet example.

According to eq.(A.5) one can obtain the maximum (absolute) value of G_2 for Gaussian (5) at the shift point $b = x_0$ as

$$max_b\, W_{G_2}(a, x_0)g = \frac{A\sigma a^{5/2}}{(a^2 + \sigma^2)^{3/2}}. \tag{6}$$

In the wavelet domain of $G_2(a, x_0)$ this dependence looks like a simple curve with one maximum. To find it, one has to solve the equation

$$\frac{d}{da} max_b\, W_{G_2}(a, x_0)g = 0$$

The corresponding calculations give the position of maximum at the scale axis is $a_{max} = \sqrt{5}\sigma$. Since the maximum location in the G_2 domain is stable when the signal is contaminated by some noise (see[24]), one can use the obtained point x_0, a_{max} to start the fit. Although the maximum of a real contaminated signal is inevitably blurred over some area in the wavelet space due to various distortions, it can nevertheless be used as a good starting point for iterations minimizing a fitting functional.

The initial signal $f(x)$, *i.e.* the $M_{\gamma\gamma}$ distributions including the background are presented in the left panels of Figs.6 and 7. Here the additional condition $E_1/E_2 > 0.8$ was applied to improve the signal/background ratio and avoid double-humped structure which will require higher orders of the expansion. The distributions look quite smooth due to the choice of a smaller step in $M_{\gamma\gamma}$ which is needed to provide more points for the wavelet analysis.

The wavelet transformation results are given in the right panels of the same Figs.6 and 7 for pC and dC, respectively. The use of WVM of the 8-th order allow us to separate noise and see a peak structure. The arrows show an approximate location of the identified peaks. Due to the trigger condition, photons from the $a \equiv \pi^0$ decay are modified shifting the distribution maximum to $M_{\gamma\gamma} \sim 125$ MeV instead of the expected 135 MeV. It is in agreement with the initial distribution

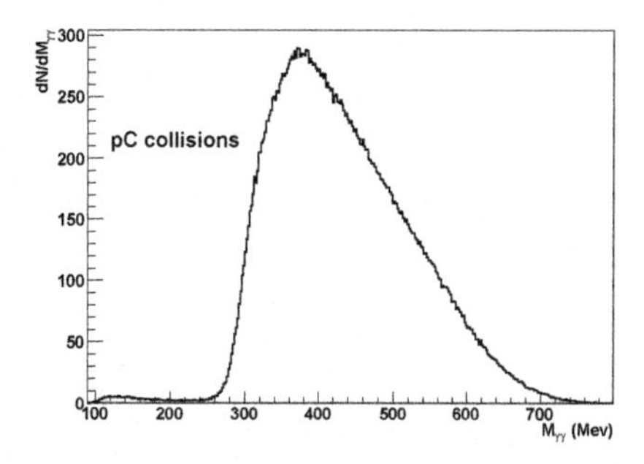

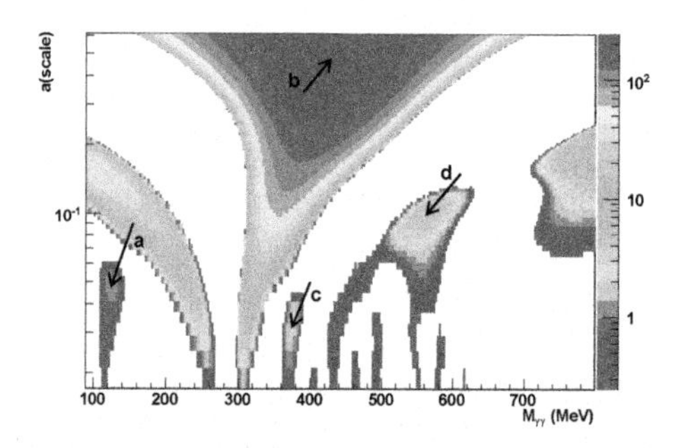

Fig. 6. The invariant mass distribution of $\gamma\gamma$ pairs from (left) and the 2D distribution of the GW of the 8-th order(right) for pC interactions. These events are selected under the same conditions as in Fig.3 but the distribution are obtained with additional condition $E_1/E_2 > 0.8$ and binning in 2 MeV.

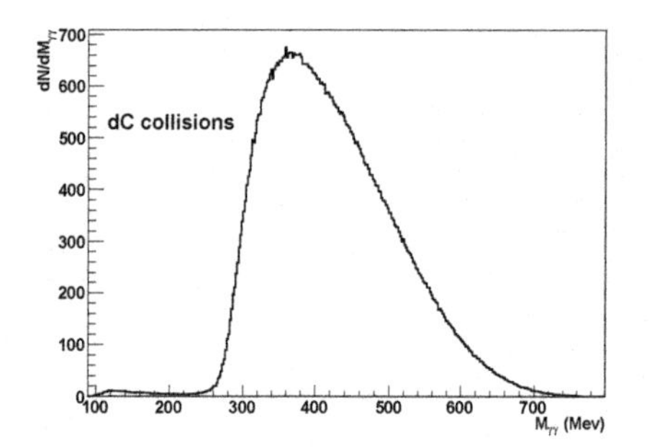

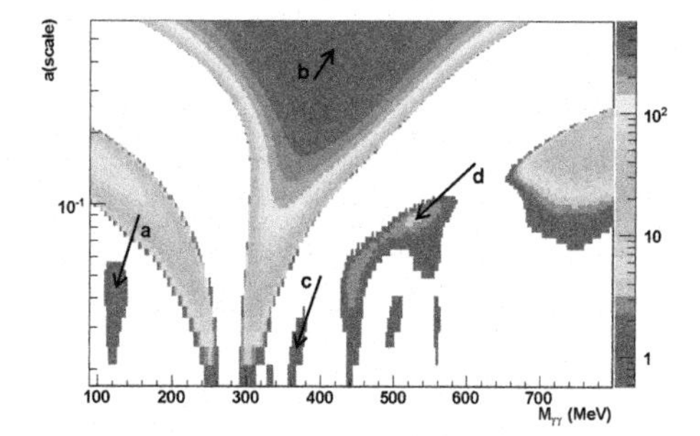

Fig. 7. The same as in Fig.6 but for dC collisions.

shown in the same figures. A huge peak of the background b dominates and its shape is quite close to the Gaussian. Photon peak from the η decay, $d \equiv \eta$, is seen quite distinctly at a proper place but its intensity is also suppressed by an additional condition $E_1/E_2 > 0.8$. In the domain of interest there is some enhancement near $M_{\gamma\gamma} \sim 370$ MeV marked in the figures $c \equiv R$. In the 2D representation this spot has not a circle shape because even in coarse binning it is not well approximated by the Gaussian (see Table.2.2) to be disturbed in wavelet space. In this respect, the c-peak in pC seems to be more pronounced than the appropriate one in dC but it follows from lower statistics in the pC case where a separate point may be better approximated by the Gaussian. Note that statistics in these cases differs by the factor of more than 3. The WVM analysis still reveals one more weak c-peak at higher $M_{\gamma\gamma}$. This is not very surprising since in coarse binning (see Fig.2) they were blurred but they are seen at a more strict selection (cf. Figs.3 – 5).

Therefore, the presented results of the continuous wavelets with vanishing momenta confirm the finding of a peak at $M_{\gamma\gamma} \sim (2-3)m_\pi$ in the $\gamma\gamma$ invariant mass distribution obtained within the standard method with the subtraction of the background from mixing events.

4. Data simulation

4.1. *About the model*

To simulate pC and dC reactions under question, we use a transport code. At high energies it is called the Quark-Gluon String Model (QGSM)[30] but at the energy of

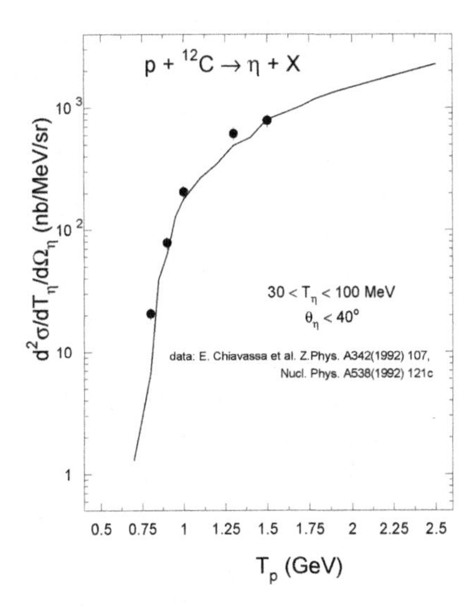

Fig. 8. The proton energy dependence of the double differential cross section for η production in pC collisions. Experimental points are from[35] .

a few GeV the string dynamics is reduced to the earlier developed Dubna cascade model[31] with upgrade elementary cross sections involved.

In the considered energy range the nuclear interactions are treated as subsequent collisions of hadrons and their resonances. The Pauli principle is involved into the model. The following γ-decay channels are taken into account: the direct decays of π^0, η, η' hadrons into two γ's, $\omega \to \pi^0\gamma$, $\Delta \to N\gamma$ and the Dalitz decay of $\eta \to \pi^+\pi^-\gamma$, $\eta \to \gamma e^+ + e^-$ and $\pi^0 \to \gamma e^+ + e^-$, the $\eta' \to \rho^0 + \gamma$, the $\Sigma \to \Lambda + \gamma$, the πN and NN-bremsstrahlung. One should note that in accordance with the recent HADES data[32] , the pn-bremsstrahlung turned out to be higher by a factor of about 5 than a standard estimate and weakly depends on the energy. This finding, being in agreement with the recent result of Ref.[33] , allowed one to resolve the old DLS puzzling[34] . This enhancement factor is included in our calculations.

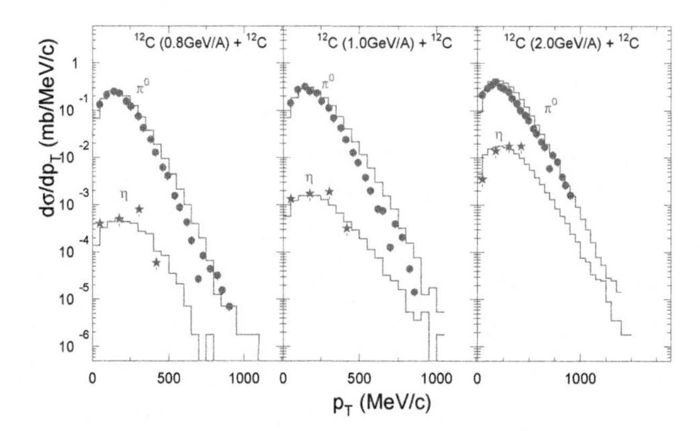

Fig. 9. Transverse momentum distributions of π^0 and η in the middle rapidity range from CC collisions at different energies. Experimental points are from the TAPS Collaboration[36] .

As a model test, in Fig.8 the excitation function for the η production is shown for the pC collisions. The model describes correctly the fast increase of the η yield near the threshold where the cross section is changed by two orders of magnitude.

The transverse momentum distributions at the mid-rapidity are presented in Fig.9 for π^0 and η produced in CC collisions at three bombarding energies. The model results are in good agreement with the TAPS experiment[36] for both neutral pions and eta mesons. So it gives us some justification for application of our model to analyze neutral particle production in the reactions considered.

4.2. *Analysis of the obtained data*

The model described above is implemented for describing the measured distributions with careful simulations of experimental acceptance. The total statistics of simulated events amounts to about 10^9 here and in every case below. As is seen from Fig.10, the model reproduces quite accurately the observed η peak in the invariant mass distribution of γ pairs but there is no enhancement in the region where experimental data exhibit a resonance-like structure which for brevity will be called below a "R-resonance". The mass of the R resonance is slightly above $2m_\pi$. As was noted, there is no particularities in this range of the phase shift diagram for the $\pi\pi$ scattering. The closest-in-mass hadron is the $f_0(600)$-meson (or the σ-meson) with the extended mass in the range of 400-1200 MeV and very large width $\Gamma = (600 - 1000)$ MeV.

To see whether such a resonance structure would be created during nuclear interaction and can survive in these strict experimental conditions, we can artificially simulate production of the R-resonance and follow its fate in the nuclear collision. It is assumed that the hypothetic R resonance can be created in every $\pi^+\pi^-$ or $\pi^0\pi^0$ interaction if its invariant mass $M_{\pi\pi}$ satisfies the Gaussian distribution with the observed parameters (see Table.2.2). The $M_{\pi\pi}$ distributions for proton- and

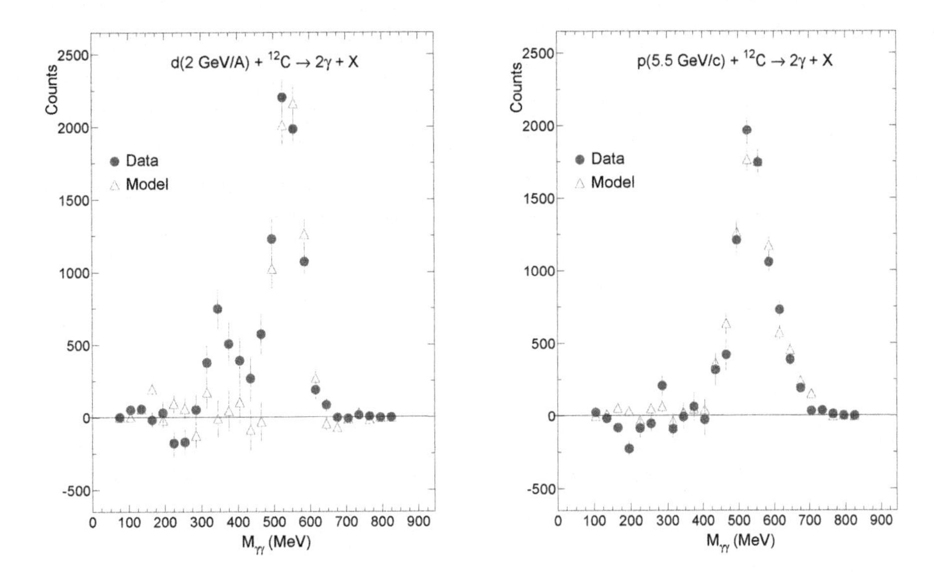

Fig. 10. The invariant mass distributions of $\gamma\gamma$ pairs from the dC (left) and pC (right) reactions after background subtraction. Both experimental (circles) and simulated (triangles) points are obtained under the same PHOTON-2 conditions.

deuteron-induced reactions is presented in Fig.11. These distributions are rather wide with a pronounced peak in the region of (2-3)m_π. The fraction of interactions identified with the formation of the R-resonance is slightly above 20 % of all $\pi\pi$ collisions. Interactions with $M_{\pi\pi} \gtrsim 650$ MeV resulting in heavy mesons ρ, ω, η come to about 5 %.

If the R resonance has been formed, it is assumed to decay immediately into two photons. This scheme can be easily realized by the Monte Carlo method within our transport model. The two-photon invariant mass spectra calculated with the inclusion of the possible R production are compared with experiment in Fig.12. Indeed, the essential part of $\gamma\gamma$ pairs survives through the strict experimental acceptance and can explain $(20-30)\%$ of the enhancement in the case of dC collisions. For pC collisions the R contribution is smaller but in agreement with experimental points. Nevertheless, one should be careful to take too seriously the estimated absolute values of the $\gamma\gamma$ pair yields from the R decay. They are obtained under extreme assumption that all R resonances decay via the two-photon channel, *i.e.* $\Gamma_R = \Gamma_{\gamma\gamma}$. However, the scalar resonance (like the σ meson) decays mainly into two pions, $\Gamma_R = \Gamma_{\pi\pi}$, and the electromagnetic decay is strongly suppressed. The two-photon decay is overwhelming ($\Gamma_R = \Gamma_{\gamma\gamma}$) only if the R mass is below the two-pion threshold. So the proposed mechanism allows one to consider properly kinematics of the $\gamma\gamma$ pairs and the role of acceptance, but it is not able to describe the absolute R yield which should be by a few orders of magnitude lower than that presented in Fig.12.

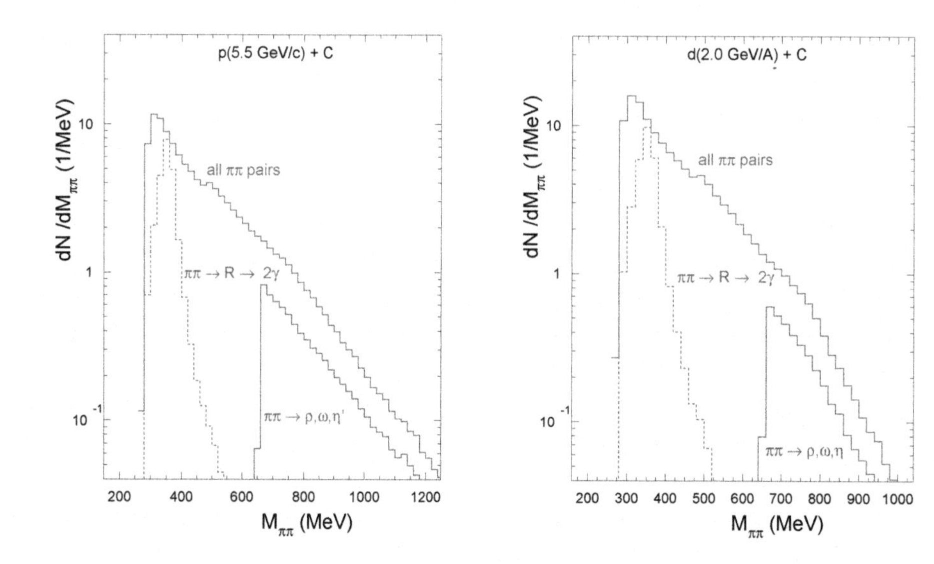

Fig. 11. Invariant mass distribution of pion pairs from $\pi\pi$ interactions in pC and dC collisions. The contributions resulting in the R-resonance and formation of ρ, ω, η' mesons are shown separately.

Model simulation allows us to disentangle different γ sources to clarify production mechanisms. This is illustrated in Fig.13. The double symbol near curves indicates the sources which both the photons came from. The π^0 decay (marked as "$\pi^0 - \pi^0$" in figures) naturally dominates in both the reactions. In the energy range considered, the pion yield rapidly increases and due to that a number of $\gamma\gamma$ pairs is higher in the pC(4.6 GeV) than in the dC(2 GeV) collisions. The η decay ("$\eta - \eta$") is seen clearly, being spread essentially due to uncertainties in the γ energy measurement. In the dC case the η maximum is more pronounced in the total distribution. It is of interest that the R resonance decay ("$R \to 2\gamma$") is also visible under PHOTON-2 conditions. A number of $\gamma\gamma$ pairs from R is higher in the pC case, but the ratios for η/R are comparable: η/R= 34.18 and 24.95 for d+C and p+C collisions, respectively. It means that a possibility to observe the R resonance depends on statistics of measured events.

One should note that since the low-mass enhancement in the invariant $\pi\pi$ spectra showed up clearly at beam energies corresponding to the excitation of Δ's in the nuclear system, the ABC effect was interpreted by a $\Delta\Delta$ excitation[37,38] . In particular, the early simplest model for ABC production in $pn \to d + X^0$ involves the excitation of both nucleons into Δ-isobars through a one-pion exchange where, after the decay of two Δ's , the final neutron-proton pair sticks together to produce the observed deuteron[37] . Though the enhancement observed in the inclusive data for the $\pi^0\pi^0$ channel turns out in some cases to be much larger than the predicted

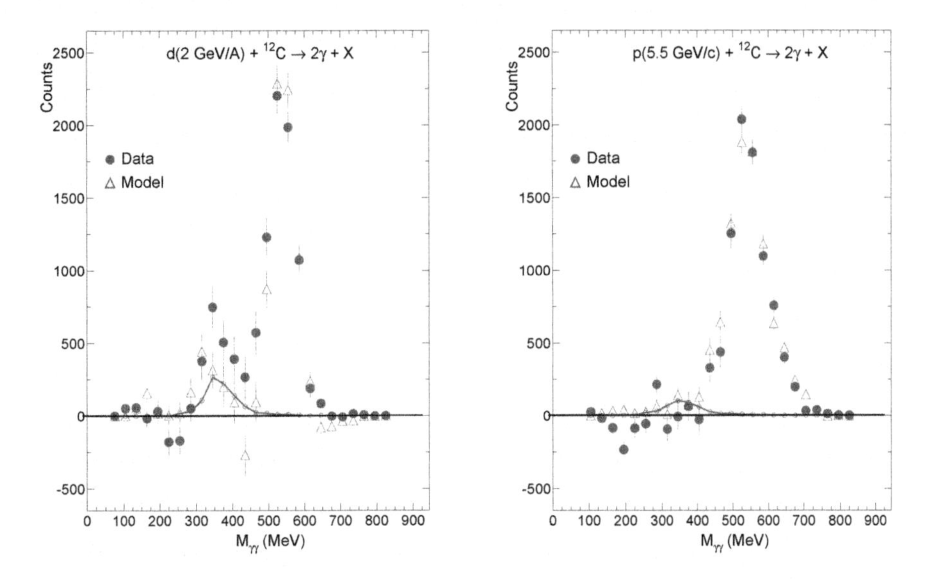

Fig. 12. Invariant mass distributions of $\gamma\gamma$ pairs from the pC and dC reactions after background subtraction. Both experimental (circles) and simulated (triangles) points are obtained under the same conditions. The contribution of photons from the R decay is shown by the solid line.

in these calculations, the $\Delta\Delta$ mechanism is still attractive. More delicate results on the vector and tensor analyzing powers in $\vec{d}d \rightarrow ^4$He+X^0 give strong quantitative support to this idea[7] .

The channel marked in Fig.13 as "$\gamma - \Delta$" corresponds to the case when photons from the Δ decay correlate with any other. Though the two-photon yield for this channel in the dC case close to that from the R decay, the maximum location is shifted to higher $M_{\gamma\gamma}$ by more than 100 MeV. If one chooses both the photons from different Δ isobars, for the PHOTON-2 conditions we have none event from 10^9 of simulated ones. One should note that we consider incoherent $\Delta\Delta$ interactions and possible attraction in this channel is not taken into account.

4.3. *The $\eta \rightarrow 3\pi^0$ decay*

As was indicated many years ago, a three-pion resonance having the mass M_η =550 MeV and isospin $T = 0$ maybe a pseudoscalar particle of positive G parity (0^{-+}). Under this assumption it was shown that the partial rates and width for this η decay will be consistent with available experimental data providing that the $\eta \rightarrow 3\pi$ channel is enhanced by a strong final state interaction[39] . This strong interaction is realized by postulating the existence of a particle having the spin and parity 0^+ and called a σ. Then the 3π decay of the η meson would proceed in two distinct steps: $\eta \rightarrow \sigma + \pi^0$, $\sigma \rightarrow 2\pi^0$ or $(\pi^+\pi^-)$ where the first step is an electromagnetic decay, while the second occurs through the strong interaction. The fit to experimental data

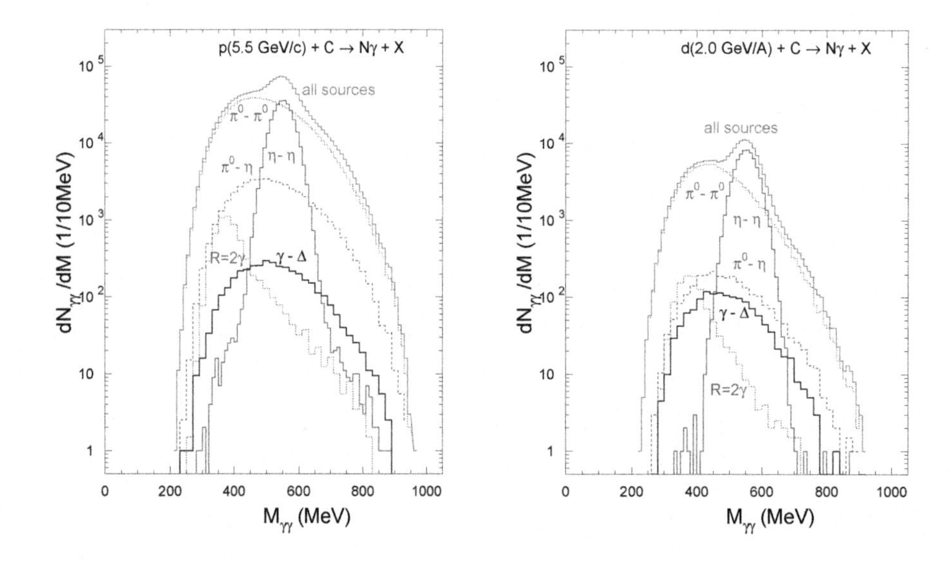

Fig. 13. The calculated $\gamma\gamma$ invariant mass distribution in pC (left) and dC (right) collisions for selected events with $N_\gamma = 2$. Contributions of different channels are shown. Symbols near curves describe sources which photons originate from.

gives the mass of a σ particle about 370 MeV and a full width of about 50 MeV[39]. These parameters coincide with those extracted from the direct analysis of pion spectra from the 3π decay of the η meson[40,41]. The enhancement of the 3π channel was argued by the presence of a strong two-pion interaction which resembles the ABC effect. It is of interest that the presence of such a pion-pion resonance improves also the calculation of the $K_1 - K_2$ mass difference[42].

To check this mechanism (see Fig.14) we simulated two channels of the η decay: the direct decay into two photons $\eta \to 2\gamma$ and $\eta \to 3\pi^0$ which then decay into photons. The last channel was calculated under two assumptions. The first, all the three pions decay independently, $\pi^0 \to \gamma\gamma$, creating 6 photons. The second version assumes that, as discussed above, two pions may interact forming σ which decay into 2 pions, so $\eta \to \sigma + \pi^0 \to 6\gamma$. As is seen (bottom panels in Fig.14), the interaction in the $\pi\pi$ channel results in some enhancement in the $\pi\pi$ invariant mass spectra as compared the non-interacting case. The $\gamma\gamma$ invariant mass spectra exhibit a spread maximum near the pion mass and practically they are identical in both the cases (dashed lines). If the PHOTON-2 experimental conditions are implemented, a clear signal (solid lines in Fig.14) at $M_{\gamma\gamma} \sim 350$ MeV appears but its intensity is by about three orders of magnitude lower than that for the η meson. It is of interest that the absolute values and shape of the $M_{\gamma\gamma}$ spectrum are again very similar in both the versions. So in the $\eta \to 3\pi^0$ decay it is hardly ever possible to disentangle the cases with and without the two-pion interaction by the detection of decay photons.

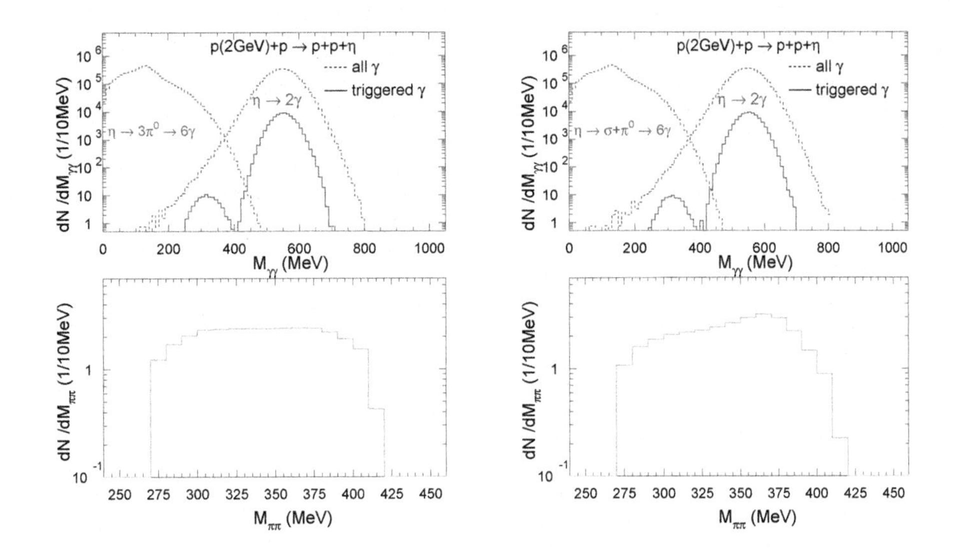

Fig. 14. The $\gamma\gamma$ (top) and $\pi\pi$ (bottom) invariant mass distributions for the η decay through the $\eta \to \gamma\gamma$ and $\eta \to 3\pi^0$ channels. The left panels correspond to the $\eta \to 3\pi^0 \to 6\gamma$ mechanism, results in the right panels include interaction of two pions via the σ meson, $\eta \to \sigma + \pi^0 \to 6\gamma$. The dashed lines are calculated for the full 4π acceptance, the solid lines take into account the PHOTON-2 experimental conditions.

4.4. *Dibaryon mechanism*

Recently, a resonance-like structure has been found by the CELSIUS-WASA Collaboration in the two-photon invariant mass spectrum near $M_{\gamma\gamma} \sim 2m_\pi$ of the exclusive reaction $pp \to pp\gamma\gamma$ at 1.2 and 1.36 GeV[43] . This observation was interpreted as the σ channel pion loops which are generated by the pp collision process and decay into the $\gamma\gamma$ channel. Interference with the underlying double bremsstrahlung background can give a reasonable account of data[43] .

In Ref.[44] , some arguments were given that such an interpretation is at least questionable and an alternative explanation was proposed where a possible origin of the structure is based on the dibaryon mechanism of the two-photon emission[45] . The proposed mechanism $NN \to d_1^\star \to NN\gamma\gamma$ proceeds trough a sequential emission of two photons, one of which is caused by production of the decoupled baryon resonance $d_1^\star$ and the other is its subsequent decay. The $pp \to pp\gamma\gamma$ transition is treated in[44] within the assumption that at a large distance the NN-decoupled six quark $d_1^\star$ state is a bound $p\Delta(1232)$ state with the spin-parity $J^P = 0^-$ and isospin $I = 2$[46] . The matrix elements were estimated phenomenologically and effects of the final state interactions between decay protons were included. This model reproduces reasonably well the experimentally observed $M_{\gamma\gamma}$ spectrum of the $pp \to pp\gamma\gamma$ reaction in the vicinity of the resonance structure[44] .

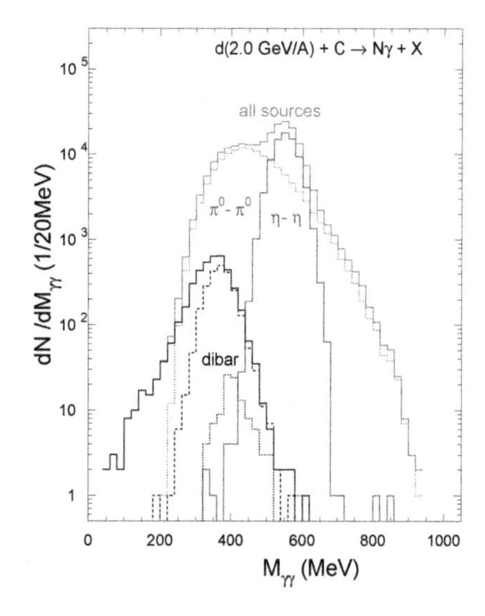

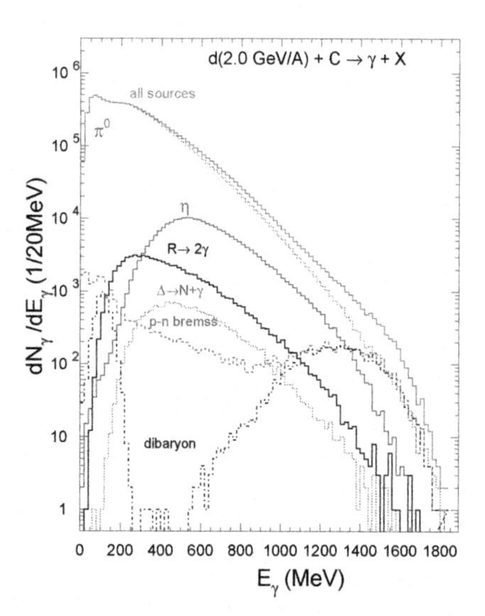

Fig. 15. The $\gamma\gamma$ invariant mass distribution (left) and energy spectra of photons (right) calculated for dC collisions with inclusion of the dibaryon mechanism. Contributions of different channels are shown similarly to Fig.13. All curves beside the dibaryon one are given for the PHOTON-2 selected events. For the dibaryon channel (marked as "dibar") the total two photon yield is presented.

We would like to check whether this dibaryon mechanism may be responsible for the peak observed in the $M_{\gamma\gamma}$ distribution of the dC collisions at $T = 2$ AGeV. The two-step scheme $NN \to d_1^\star\gamma \to NN\gamma\gamma$, where the dibaryon mass is $m_d = m_N + m_\Delta = 2.182$ GeV, can be easily simulated and included into our transport model. The only unknown quantity is the cross section of this process. We estimated it by means of the linear extrapolation of the two available points measured at 1.2 and 1.36 GeV[43] till the energy of about 2 GeV.

In Fig.15, such model calculation results with inclusion of the dibaryon channel are presented. If the beam energy of the pp collision is fixed by 2 GeV, the photon energy at the first step $pp \to d_1^\star\gamma$ will be a line, $E_{\gamma 1} = (s - M_d^2)/(2\sqrt{s}) = 640$ MeV in the c.m.s. system. Here s is the total pp colliding energy squared. In the lab. system the maximal photon energy reaches about 1.5 GeV. However, the photon from the second stage will be soft, being defined by the baryon mass. Nevertheless, the maximum position of appropriate two photon distribution moves also with the energy increase. The total photon spectra from dibaryon and other sources from dC interactions are shown in the right panel of Fig.15 and it has two-bump structure which mirrors the two-step production mechanism. One should note that in contrast with all other results in this figure the distributions of the dibaryon channel are obtained for the full 4π acceptance without any limitation on energy. If the PHOTON-2 selection conditions are implemented to the dibaryon channel, the low energy photons are cut and we get no two-photon pair from the dibaryon among

10^9 simulated collisions. Therefore, this dibaryon mechanism cannot explain the observed anomaly.

5. Estimate of the η and R production cross sections and resonance widths

The summed number of pC- and dC-interactions in the experiment amounts to $\sim 3 \cdot 10^{11}$ and $\sim 2 \cdot 10^{12}$, respectively. The inelastic cross sections of the observed pC and dC reactions are $\sigma_{inel}(pC)$ =411 mb and $\sigma_{inel}(dC)$ = 426 mb[47] , respectively.

The cross section for the η production in dC collisions (similarly for pC interactions) is defined as follows:

$$\sigma(dC \to \eta + X) = \sigma_{inel}(dC) \cdot \frac{N_{\eta}^{exp}}{N_{dC-inter}} \cdot \frac{N_{all\eta}^{mod}}{N_{\eta}^{mod}/K_{opt}} . \qquad (7)$$

Here the first two factors are the reaction cross section and the measured mean multiplicity of η mesons. A number of true dC interactions resulting in the η production is given as

$$N_{dC-inter} = K_{empty} \cdot K_{beam-absorb} \cdot N_d \ , \qquad (8)$$

where the total number of beam particles passing through the target $N_d \sim 2.2\ 10^{12}$ is corrected on possible interactions outside the target $K_{empty} \sim 0.05$, to be estimated by a special experiment with an empty target and the beam absorption $K_{beam-absorb} = 0.8 \pm 0.1$. The last factor mirrors a particularity of internal beam experiments where the direct monitoring is impossible and a part of the initial beam does not interact during the working circle.

The third factor in eq.(7) is the ratio of the total number of simulated η mesons to a number of η's decaying into two photons under PHOTON-2 experimental conditions. The coefficient $K_{opt} = 8.93$ takes into account the rotation of the modelled events in the ϕ plane to find other possible $\gamma\gamma$ pairs in the given event satisfying the selection conditions. Assuming that photons in the event are distributed homogeneously in ϕ, this trick allows one to increase effectively statistics of the selected events.

So, for the η production in dC collisions we have

$$\sigma(dC \to \eta + X) = 1.31 \pm 0.11^{+1.24}_{-0.89}\ mb\ .$$

For the case of pC collisions we get

$$\sigma(pC \to \eta + X) = 3.2 \pm 0.2^{+4.5}_{-1.9}\ mb\ .$$

If the cross section for the η production is known, the cross section for the

R-resonance may be estimated as follows:

$$\sigma(dC \to R \to \gamma\gamma) = \sigma(dC \to \eta + X) \cdot Br(\eta \to \gamma\gamma) \cdot \cdot \frac{N^{exp}(R \to \gamma\gamma)/\epsilon_R}{N^{exp}(\eta \to \gamma\gamma)/\epsilon_\eta}$$

$$= (0.075 \pm 0.018) \cdot \sigma(dC \to \eta + X) = 98 \pm 24^{+93}_{-67}\ \mu b\ , \tag{9}$$

where the branching ratio is $Br(\eta \to \gamma\gamma) = 0.38$, $\epsilon_R = N^{mod}(R \to \gamma\gamma)/N^{mod}_{tot}(R)$ and $\epsilon_\eta = N^{mod}(\eta \to \gamma\gamma)/N^{mod}_{tot}(\eta)$ are the registration and selection efficiency. The measured average reduced multiplicities are compared in Fig.16 with available systematics for meson production near the threshold energies. This simple energy

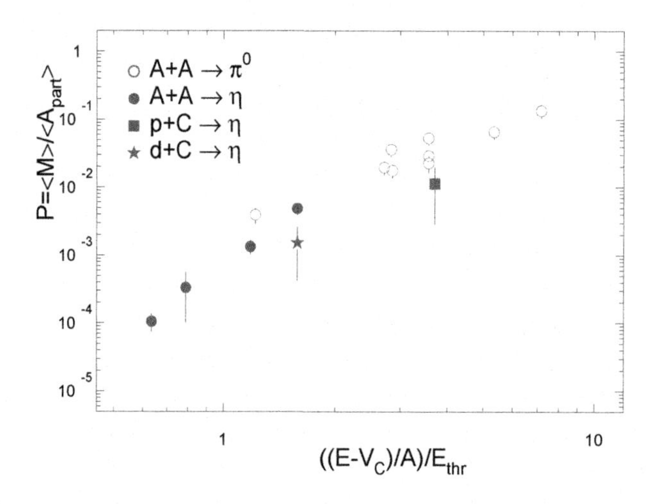

Fig. 16. Meson production probability per participant nucleon as a function of the bombarding energy per nucleon corrected to the Coulomb barrier V_c and reduced to the respective meson production threshold E_{thr}. Experimental points of the TAPS collaboration are taken from the review article[48] . Our points for dC and pC collisions are plotted by the star and filled square.

scaling systematics for the subthreshold and near threshold particle production was proposed in[49] . A number of participant nucleons is defined as

$$A_{part} = \frac{A_p\ A_T^{2/3} + A_T\ A_p^{2/3}}{A_p^{1/3} + A_T^{1/3}}\ . \tag{10}$$

The bombarding energy is corrected for the Coulomb barrier V_c. This systematics is valid also for K and ρ mesons[48,49] . As is seen, the measured η production points are somewhat below the general trend. Partially, it may be caused by the point that all other experimental points correspond to heavier nuclear systems and eq.(10) is not very justified for our reactions.

As to the true internal width w of the observed resonances, they are defined by the measured width w_{measur} and also specified by the spectrometer resolution w_{sp} :

$$w = (w^2_{measur} - w^2_{sp})^{1/2} \ . \quad (11)$$

The spectrometer resolution depends on the $M_{\gamma\gamma}$ range. For the R and η invariant mass range we have, respectively,

$$2w_{app}(340 < M_{\gamma\gamma} < 360MeV) = 52.6 \ MeV,$$
$$2w_{app}(540 < M_{\gamma\gamma} < 560MeV) = 68.6 \ MeV. \quad (12)$$

So, according to eq.(11) the intrinsic widths of detected resonances are

$$w(\eta \to \gamma\gamma) \approx 0.$$
$$w(R \to \gamma\gamma) \simeq 49.2 \pm 18.6 \ . \quad (13)$$

As is expected, the width of the η-meson practically equals 0, whereas it essentially differs from zero in the observed resonance. The value of $2w$ in the Gaussian distribution (1) practically coincides with the width Γ in the Breit-Wigner function; thus, the intrinsic width of the observed resonance structure is about 49 ± 19 MeV.

6. Conclusions

Thus, based on a thorough analysis of experimental data measured at the JINR Nuclotron and record statistics of 2339 ± 340 events of $1.5 \cdot 10^6$ triggered interactions of a total number $2 \cdot 10^{12}$ of dC-interactions there was observed a resonance-like enhancement at the mass $M_{\gamma\gamma} = 360 \pm 7 \pm 9$ MeV, width $\Gamma = 49 \pm 19$ MeV, and preliminary production cross section $\sigma_{\gamma\gamma} \sim 98 \ \mu b$ in the invariant mass spectrum of two photons produced in dC-interactions at momentum of incident deuterons 2.75 GeV/c per nucleon. A structure like this is not observed in the $M_{\gamma\gamma}$ spectrum from pC (5.5 GeV/c) interactions while the η meson is clearly seen in both the cases. These results, obtained by means of the mixing event background, are confirmed by the wavelet analysis.

To certain extent this enhancement at $M_{\gamma\gamma} \sim (2-3)m_\pi$ is similar to the puzzling ABS effect observed for two-pion pairs from nucleon-nucleon and lightest nuclei collisions at the near threshold energy. In the given work we see it in the $\gamma\gamma$ channel and measurements are extended to a heavier system.

To understand the origin of the observed structure, several dynamic mechanisms were attempted: production of the hypothetic R resonance in $\pi\pi$ interactions during the evolution of the nuclear collision, formation of the R resonance with participation of photons from the Δ decay, the $\pi^0\pi^0$ interaction effect in the $3\pi^0$ channel of the η decay, a particular decoupled dibaryon mechanism. Unfortunately, none of these mechanisms is able to explain the measured value of the resonance-like enhancement, though they contribute to the invariant mass region in question.

To verify the above conclusions and determine more accurately the mass, width and cross section of the observed resonance structure, new experiments are required to be carried out under conditions appropriate for registration of pairs of two photons within the invariant mass interval of 300-400 MeV.

Acknowledgments

We are grateful to A.S. Danagulyan, S.B. Gerasimov, V.D. Kekelidze, A.S. Khrykin, E.E. Kolometsev, A.M. Sirunyan, O.V. Teryaev, G.A. Vartapetyan for numerous fruitful discussions. We thank S.V. Afanasev, V.V. Arkhipov, A.S. Artemov, A.F. Elishev, A.D. Kovalenko, V.A. Krasnov, A.G. Litvinenko, A.I. Malakhov, G.L. Melkumov, S.N. Plyashkevich and the staff of the Nuclotron for their help in conducting the experiment, as well as B.V. Batyunya, A.V. Belozerov, A.G. Fedunov, V.V. Skokov and V.D. Yudichev for their help in analyzing the data.

The work was supported in part by the Russian Foundation for Basic Research, grant 08-02-01003- and a special program of the Ministry of Education and Science of the Russian Federation, grant RNP.2.1.1.5409.

Appendix A. Continuous wavelets with vanishing momenta

The family of continuous wavelets with *vanishing momenta* (VMW) is presented here by Gaussian wavelets (GW) which are generated by derivatives of the Gaussian function (5). For canonical Gaussian with $x_0 = 0$; $\sigma = 1$ and $A = 1$ one obtains

$$g_n(x) = (-1)^{n+1} \frac{d^n}{dx^n} e^{-x^2/2}, \tag{A.1}$$

where $n > 0$ is the order of the $g_n(x)$ wavelet. The normalizing coefficients of these wavelets C_{g_n} are $2\pi(n-1)!$.

The most known in the GW family is the second order GW

$$g_2(x) = (1 - x^2)\mathrm{e}^{-\frac{x^2}{2}},$$

which is also known as "the Mexican hat"[23] .

We use here also GW of higher orders, in particular,

$$g_4(x) = (6x^2 - x^4 - 3)\mathrm{e}^{-\frac{x^2}{2}} \tag{A.2}$$

$$g_6(x) = (x^6 - 15x^4 + 45x^2 - 15)\mathrm{e}^{-\frac{x^2}{2}} \tag{A.3}$$

$$g_8(x) = (x^8 - 28x^6 + 240x^4 - 420x^2 + 105)\mathrm{e}^{-\frac{x^2}{2}} \tag{A.4}$$

It is a remarkable fact that the wavelet transformation of Gaussian (5) looks as the corresponding wavelet. Therefore the general expression for the n-th wavelet coefficient has the following form (see its derivations in[24]):

$$W_{g_n}(a,b)g = \frac{A\sigma a^{n+1/2}}{\sqrt{(n-1)!}\ s^{n+1}}\, g_n\left(\frac{b - x_0}{s}\right), \tag{A.5}$$

where we denote $s = \sqrt{a^2 + \sigma^2}$.

One of the attractive features of the wavelet based estimations is their remarkable robustness to background noise, binning range, and data missing (see[24]).

The parameters a and b of continuous wavelets in eq.(2) are changing continuously, which leads to the redundant representation of the data. In some cases, the above mentioned GW properties are quite useful, in particular, this redundant representation facilitates careful spectrum manipulations. The price of this redundancy consists in slow speed of calculations. Besides, all signals to be analyzed have in practice a discrete structure.

It is noteworthy that GW should be used with some care since they are nonorthogonal, which may disturb amplitudes of the filtered signal after their inverse transform. In this respect, the discrete wavelet transform looks more preferable for many applications of computing calculations with real data[25] and deserves special consideration. As was noted above, in our particular case of the continuous VMW one can identify resonances without the inverse transformation.

References

1. A. Abashian, N. E. Booth and K. M. Crowe, *Phys. Rev. Lett.* **5**, 3258 (1960); N. E. Booth, A. Abashian and K. M. Crowe, *Phys. Rev. Lett.* **7**, 35 (1961); *Phys. Rev.* **132**, 2309 (1963).
2. J. Banaigs et al., *Nucl. Phys. B* **67**, 1 (1973).
3. R. J. Homer et al., *Phys. Lett.* **9**, 72 (1964).
4. J. H. Hall et al., *Nucl. Phys. B* **12**, 573 (1969).
5. I. Bar-Nir et al., *Nucl. Phys. B* **54**, 17 (1973).
6. J. Banaigs et al., *Nucl. Phys. B* **105**, 52 (1976).
7. R. Wurzinger et al., *Phys. Lett. B* **445**, 423 (1999).
8. F. Plouin et al., *Nucl. Phys. B* **302**, 413 (1978).
9. A. Abdivaliev at al., *Sov. J. Nucl. Phys.* **29**, 796 (1979); *Nucl. Phys. B* **168**, 385 (1980).
10. F. Plouin, P. Fleury and C. Wilkin, *Phys. Rev. Lett.* **65**, 690 (1990).
11. B. Richter, *Phys. Rev. Lett.* **9**, 217 (1962).
12. R. Del Fabbro et al., *Phys. Rev. Lett.* **12**, 674 (1964).
13. A. Codino and F. Plouin, *Production of light mesons and multipion systems in light nuclei interactions*, LNS/Ph/94-06.
14. T. Skorodko et al., CELSIUS-WASA Collaboration, arXiv:nucl-ex/0612016.
15. H. Clement et al., CELSIUS-WASA Collaboration, arXiv:0712.4125.
16. J. Yonnet et al., *Phys. Rev. C* **63**, 014001 (2001).
17. Kh. U. Abraamyan et al., in *Proc. of the Int. Workshop "Relativistic Nuclear Physics: from Hundreds MeV to TeV", Dubna*, 228 (2005); *Workshop Round Table Discussion: "Searching for the mixed phase of strongly interacting matter at the JINR Nuclotron", JINR, Dubna, 7 - 9 July 2005*, http://theor.jinr.ru/meetings/2005/roundtable/.
18. Kh. U. Abraamyan, A. N. Sissakian and A. S. Sorin, Observation of new resonance structure in the invariant mass spectrum of two gamma-quanta in dC-interactiona at momentum 2.75 GeV/c per nucleon, nucl-ex/0607027.
19. Kh. U. Abraamyan et al., *Phys. Lett. B* **323**, 1 (1994); *Yad. Fiz.* **59**, 271 (1996); **60**, 2014 (1997); **68**, 1020 (2005).
20. Kh. U. Abraamyan et al., *PTE* **1**, 57 (1989); **6**, 5 (1996).
21. M. N. Khachaturian, in *JINR Communication E1-85-55, Dubna, 1985*; R. G. Astvatsaturov et. al., *Nucl. Instr. Methods* **163**, 343 (1979).
22. I. Daubechies, *IEEE Trans. Inform. Theory* **36**, 961 (1990).
23. *Proc. of Symp. in Appl. Math.*, ed. by I. Daubechies.

24. G. Ososkov and A. Shitov, *Comp. Phys. Comm* **126/1-2**, 149 (2000).
25. S. A. Mallat, Theory for multi-resolution signal decomposition: the wavelet representation. *IEEE Trans. Pattern Analysis and Machine Intelligence* (1989).
26. G. A. Ososkov and A. V. Stadnik, Wavelet approach for peak finding in heavy ion physica, http://www.gsi.de/documents/DOC-2007-Oct-101-1.pdf
27. I. Dremin, O. Ivanov and V. Nechitailo, *Uspekhi Fiz. Nauk* **171**, 465 (2001) (in Russian).
28. N. Astafieva, *Uspekhi Fiz. Nauk* **166**, 1145 (1996) (in Russian).
29. T. S. Belozerova, P. G. Frick and V. K. Henner, *Yad. Fiz.* **66** (2003) (translated as Phys. Atom. Nucl. **66**, 1269 (2003)).
30. N. S. Amelin, K. K. Gudima and V. D. Toneev, *Sov. Journ. of Nuclear Phys.* **51**, 1730 (1990); N. S. Amelin et al., *Sov. Journ. of Nuclear Phys.* **52**, 172 (1990); *Phys. Rev. C* **44**, 1541 (1991); V. D. Toneev et al., *Nucl.Phys. A* **519**, 463c (1990); N. S. Amelin, et al., *Phys. Rev. Letters* **67**, 1523 (1991);. V. D. Toneev et al, *Sov. Jour. of Nucl. Phys.* **54**, 1272 (1992); N. S. Amelin et al., *Phys. Rev. C* **47**, 2299 (1993).
31. K. K. Gudima and V. D. Toneev. *Nucl. Phys. A* **400**, 173c (1983); K. K. Gudima et al, *Nucl. Phys. A* **401**, 329 (1983).
32. G. Agakishiev et al., HADES Collaboration, *Phys. Rev. Lett.* **98**, 052302 (2007); arXiv:0711.4281 .
33. L. P. Kaptari and B. Kämpfer, *Nucl. Phys. A* **764**, 338 (2006).
34. E. L. Bratkovskaya and W. Cassing, arXiv:0712.0635
35. E. Chiavassa et al., *Z. Phys. A* **342**, 107 (1992); *Nucl. Phys. A* **538**, 121c (1992).
36. Averbeck et al., TAPS Collaboration, *Z. Phys. A* **359**, 65 (1997).
37. *Phys. Lett. B* **43**, 68 (1973); I. Bar-Nir, T. Risser and M.D. Shuster, *Nucl. Phys. B* **B87**, 109 (1975); J. C Anjos, D. Levy and A. Santoro, **B67**, 109 (1973).
38. A. Gardestig, C. Feld and C. Wilkin, *Phys. Rev. C* **59**, 6208 (1999); *Phys. Lett.* **421**, 41 (1998).
39. L. M. Brown and P. Singer, *Phys. Rev. Lett.* **8**, 460 (1962).
40. F.S. Crawford, Jr. et al., *Phys. Rev. Lett.* **11**, 564 (1963).
41. A. Roy for CBELSA TAPS Collaboration, *PRAMANA - J. of Physics* **66**, 921 (2006).
42. K. Nishijima, *Phys. Rev. Lett.* **12**, 39 (1964).
43. M. Bashkhanov et al., *Int. J. of Mod. Phys. A* **20**, 554 (2005); arXiv:hep-exp/0406081.
44. A. S. Krykhin and S. B. Gerasimov, arXiv:0710.3331 .
45. A. S. Khrikin et al., *Phys. Rev. C* **64**, 034002 (2001); *Nucl. Phys. A* **721**, 625c (2003).
46. A. Matsuyama, *Phys. Lett. B* **408**, 25 (1997).
47. V. S. Barashenkov, Cross-sections of interactions of particles and nuclei with nuclei. Dubna, 1993.
48. W. Kuhn, *Czech. J. Phys.* **4**, 537 (1995).
49. V. Metag, *Prog. Part. Nucl. Phys.* **30**, 75 (1993).

CORRELATION RADII BY FAST HADRON FREEZE-OUT GENERATOR

Iu. A. KARPENKO[1], R. LEDNICKY [2,4], I. P. LOKHTIN [3] , L. V. MALININA[2,3], Yu. M. SINYUKOV[1] and A. M. SNIGIREV[3]

(1) Bogolyubov Institute for Theoretical Physics, Kiev, 03143, Ukraine;
(2) Joint Institute for Nuclear Research, Dubna, Moscow Region, 141980, Russia;
(3) M.V. Lomonosov Moscow State University, D.V. Skobeltsyn Institute of Nuclear Physics, 119992, Moscow, Russia;
(4) Institute of Physics ASCR, Prague, 18221, Czech Republic
E-mail: malinina@lav01.sinp.msu.ru

The predictions for correlation radii in the central Pb+Pb collisions for LHC $\sqrt{s}_{NN} =$ 5500 GeV are given in the frame of FAST HADRON FREEZE-OUT GENERATOR (FASTMC).

One of the most spectacular features of the RHIC data, refereed as "RHIC puzzle", is the impossibility to describe simultaneously momentum-space measurements and the freeze-out coordinate-space ones (femtoscopy) by the existing hydrodynamic and cascade models or their hybrids. However, a good description of SPS and RHIC data have been obtained in various models based on hydro-inspired parametrizations of freeze-out hypersurface. Thus, we have achieved this goal within our fast hadron freeze-out MC generator (FASTMC)[1] . In FASTMC, particle multiplicities are determined based on the concept of chemical freeze-out. Particles and hadronic resonances are generated on the thermal freeze-out hypersurface, the hadronic composition at this stage is defined by the parameters of the system at chemical freeze-out[1] . The input parameters which control the execution of our MC hadron generator in the case of Bjorken-like parameterization of the thermal freeze-out hypersurface (similar to the well known "Blast-Wave" parametrization with the transverse flow) for central collisions are the following: temperature T^{ch} and chemical potentials per a unit charge $\widetilde{\mu}_B, \widetilde{\mu}_S, \widetilde{\mu}_Q$ at chemical freeze-out, temperature T^{th} at thermal freeze-out, the fireball transverse radius R, the mean freeze-out proper time τ and its standard deviation $\Delta\tau$ (emission duration), the maximal transverse flow rapidity $\rho_u^{\max}$. We considered here the naive "scaling" of the existing physical picture of heavy ion interactions over two orders of magnitude in $\sqrt{s}$ to the maximal LHC energy $\sqrt{s}_{NN} = 5500$ GeV. The model parameters obtained by the fitting within FASTMC generator of the existing experimental data on m_t-spectra, particle ratios, rapidity density dN/dy, k_t-dependence of the correlation radii $R_{\mathrm{out}}, R_{\mathrm{side}}, R_{\mathrm{long}}$ from SPS ($\sqrt{s}_{NN} = 8.7 - 17.3$ GeV) to RHIC ($\sqrt{s}_{NN} = 200$ GeV) are shown in Fig. 1. For

LHC energies we have fixed the thermodynamic parameters at chemical freeze-out as the asymptotic ones: $T^{\rm ch} = 170$ MeV, $\widetilde{\mu}_B = \widetilde{\mu}_S = \widetilde{\mu}_Q$=0 MeV. The linear extrapolation of the model parameters in $\log(\sqrt{s})$ to LHC ($\sqrt{s}_{NN} = 5500$ GeV) is shown in Fig. 1 by open symbols. The extrapolated values are the following: $R \sim 11$ fm, $\tau \sim 10$ fm/c, $\Delta\tau \sim 3.0$ fm/c, $\rho_u^{\rm max} \sim 1.0$, $T^{\rm th} \sim 130$ MeV. The density of charged particles at mid-rapidity obtained with these parameters is $dN/dy = 1400$, i.e. twice larger than at RHIC $\sqrt{s}_{NN} = 200$ GeV in coincidence with the naive extrapolation of dN/dy. These parameters yield only a small increase of the correlation radii $R_{\rm out}, R_{\rm side}, R_{\rm long}$ (Fig. 2).

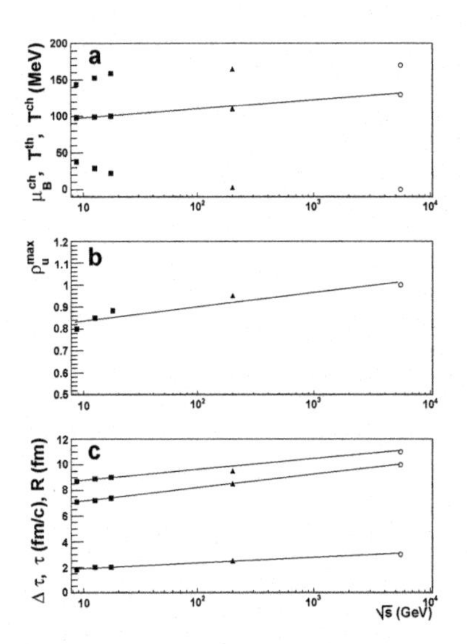

Fig. 1. FASTMC parameters versus $\log(\sqrt{s})$ for SPS $\sqrt{s} = 8.7 - 17.3$ GeV (black squares), RHIC $\sqrt{s} = 200$ GeV (black triangles) and LHC $\sqrt{s} = 5500$ GeV(open circles): (**a**) $T^{\rm ch}$, $T^{\rm th}$, μ^B, (**b**) ρ_u^{max}, (**c**) τ, R and $\Delta\tau$.

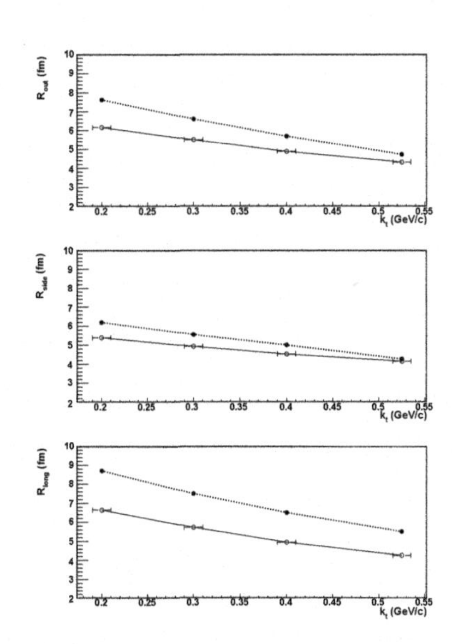

Fig. 2. The $\pi^+\pi^+$ correlation radii in longitudinally comoving system at mid-rapidity in central Au+Au collisions at $\sqrt{s}_{NN} =$ 200 GeV from the STAR experiment[2] (open circles) and the FASTMC calculations for LHC $\sqrt{s} = 5500$ GeV (black squares).

References

1. N. S. Amelin, R. Lednicky, T. A. Pocheptsov, I. P Lokhtin, L. V. Malinina, A. M. Snigirev, Iu. A. Karpenko and Yu. M. Sinyukov, *Phys. Rev. C* **74** 064901 (2006).
2. J. Adams *et al.* (STAR Collaboration) *Phys. Rev. C* **71** 044906 (2005).

HIGH-p_T FEATURES OF z-SCALING AT RHIC AND TEVATRON

M. V. TOKAREV[1,♮], I. ZBOROVSKY[2,*],
T. G. DEDOVICH[1]

(1) Joint Institute for Nuclear Research,
Joliot Curie st., 6, Dubna, Moscow region, 141980 Russia;
(2) Nuclear Physics Institute ASCR, Řež, Czech Republic
♮ *E-mail: tokarev@sunhe.jinr.ru ,*
* *E-mail: zborovsky@ujf.cas.cz*

Experimental data on inclusive cross sections of jet, direct photon, and high-p_T hadron production in $pp/\bar{p}p$ and AA collisions are analyzed in the framework of z-scaling. The analysis is performed with data obtained at ISR, S$\bar{p}$pS, RHIC, and Tevatron. Scaling properties of z-presentation of the inclusive spectra are verified. Physical interpretation of the variable z and the scaling function $\psi(z)$ is discussed. We argue that general principles of self-similarity, locality, and fractality reflect the structure of the colliding objects, interaction of their constituents, and particle formation at small scales. The obtained results suggest that the z-scaling may be used as a tool for searching for new physics phenomena beyond Standard Model in hadron and nucleus collisions at high transverse momentum and high multiplicity at U70, RHIC, Tevatron, and LHC.

1. Introduction

Study of particle production at large transverse momenta in high energy collisions of hadrons and nuclei is of interest to search for exotic phenomena such as quark compositeness[1] , extra dimensions[2] , black holes[3] , fractal space-time[4] , and collective phenomena such as phase transition of nuclear matter and formation of quark-gluon plasma[5] . New experimentally established features of particle formation could be crucial for precise test of Quantum chromodynamics (QCD) and Electroweak (EW) theories both in perturbative and non-perturbative regimes. Many phenomenological approaches were suggested for description of regularities reflecting general properties (locality, self-similarity) of hadron and nucleus interactions at a constituent level[6−15] .

In the present paper we use the concept of the z-scaling[16−20] for analysis of experimental data on inclusive spectra of hadron, direct photon, and jet production in $pp/\bar{p}p$ and AA collisions at RHIC and Tevatron. The procedure for construction of the scaling variable z and scaling function $\psi(z)$ for different types of the produced objects is sufficiently simple. The construction is expressed via the experimentally measured inclusive cross section $Ed^3\sigma/dp^3$, the multiplicity density $dN_{\rm ch}/d\eta$, and kinematical characteristics of colliding and produced particles. Scaling property is used to predict particle spectra of $J/\psi, D^0, B^+$ mesons and Z, W^+ bosons at higher collision energies and transverse momenta. We suggest to use the z-scaling as a tool for searching for new physics phenomena of particle production in high transverse momentum and high multiplicity region at U70, Tevatron, RHIC, and LHC.

2. z-Scaling

We would like to remind basic ideas and some formulae of the developed approach[16–20] which concern approximations used in the present analysis of experimental data. The z-scaling is based on the assumption[11] that gross features of the inclusive particle distribution of the inclusive reaction

$$M_1 + M_2 \to m_1 + X \tag{1}$$

can be expressed at high energies in terms of the kinematic characteristics of the corresponding constituent subprocess

$$(x_1 M_1) + (x_2 M_2) \to m_1/y + (x_1 M_1 + x_2 M_2 + m_2/y). \tag{2}$$

Here x_1 and x_2 are the fractions of the 4-momenta P_1 and P_2 of the incoming objects with the masses M_1 and M_2 carried by the interacting constituents. The inclusive particle with the mass m_1 and the 4-momentum p carries the fraction y of the 4-momentum of the outgoing constituent. The parameter m_2 is introduced to satisfy the internal conservation laws (for baryon number, isospin, strangeness, and so on). Its value is determined from the corresponding exclusive reaction. For example, the parameter m_2 was set to $m_n - m_p$ for the exclusive process $p + p \to \pi^+ + (p + n)$ as follows from the relation $\pi^+ + (p + n) = \pi^+ + p + p + (n - p)$ satisfying to the baryon and electric charge conservation laws. For other particles we determine m_2 in a similar way. It is assumed that the constituent interaction satisfies the energy-momentum conservation law and is subject to the condition

$$(x_1 P_1 + x_2 P_2 - p/y)^2 = (x_1 M_1 + x_2 M_2 + m_2/y)^2. \tag{3}$$

The equation is expression of the locality of hadron interactions at constituent level.

The scaling variable z is defined as follows[20] * .

$$z = \frac{s_\perp^{1/2}}{(dN_{ch}/d\eta|_0)^c \cdot m_0} \cdot \Omega^{-1}. \tag{4}$$

The symbol $s_\perp^{1/2}$ denotes minimal transverse kinetic energy of the underlying constituent sub-process. It consists of two parts $s_\lambda^{1/2}$ and $s_\chi^{1/2}$ which represent the energy for creation of the inclusive particle and its recoil, respectively. The quantity $dN_{ch}/d\eta|_0$ is multiplicity density of charged particles in the central interaction region at (pseudo)rapidity $\eta = 0$. The parameter c has physical meaning of a "specific heat" of the produced medium. The mass constant m_0 is fixed at the value of the nucleon mass. The quantity

$$\Omega(x_1, x_2, y) = (1 - x_1)^{\delta_1}(1 - x_2)^{\delta_2}(1 - y)^{\epsilon}. \tag{5}$$

---

*Other modifications of z are described in[16,21,22]

is function of the momentum fractions x_1, x_2, and y and depends on the parameters δ_1, δ_2, and ϵ interpreted as fractal dimensions in space of the momentum fractions. It is proportional to relative number of all parton configurations containing the incoming constituents which carry the fractions x_1 and x_2 of the momenta P_1 and P_2 and the outgoing constituent which fraction y is carried by the inclusive particle with the momentum p. The momentum fractions are determined in a way to minimize the function $\Omega^{-1}(x_1, x_2, y)$ taking into account the energy-momentum conservation law in the binary collision with the condition (3). This is equivalent to the solution of the system of nonlinear equations

$$\partial\Omega(x_1, x_2, y)/\partial x_1 = 0, \quad \partial\Omega(x_1, x_2, y)/\partial x_2 = 0, \quad \partial\Omega(x_1, x_2, y)/\partial y = 0 \qquad (6)$$

where x_1, x_2, and y are coupled by Eq. (3).

The scaling function $\psi(z)$ is expressed in terms of the experimentally measured inclusive invariant cross section $Ed^3\sigma/dp^3$, the multiplicity density $dN/d\eta$, total inelastic cross section σ_{in} and the kinematical variables (masses and momenta) characterizing the inclusive reaction. It can be written as follows

$$\psi(z) = -\frac{\pi s}{(dN/d\eta)\sigma_{in}} J^{-1} E \frac{d^3\sigma}{dp^3}. \qquad (7)$$

Here s is the collision center-of-mass energy squared, J is the corresponding Jacobian, and $dN/d\eta$ is multiplicity density of respective particle species. The function $\psi(z)$ determined by Eq. (7) satisfies to the normalization condition

$$\int_0^\infty \psi(z) dz = 1. \qquad (8)$$

The relation allows us to interpret $\psi(z)$ as a probability density to produce an inclusive particle with the corresponding value of the variable z.

3. *z*-Scaling in *pp* and *AA* collisions at RHIC

In this section we analyze experimental data on particle transverse momentum spectra obtained in *pp* and *AA* collisions at RHIC.

3.1. π^0 *mesons*

The PHENIX and STAR collaborations published the new data[23,24] on the inclusive cross sections of π^0-mesons produced in *pp* collisions in the central rapidity range at the energy $\sqrt{s} = 200$ GeV. The transverse momenta of π^0-mesons were measured up to $p_T \simeq 20$ GeV/c. The p_T and z-presentations of these data as well as data obtained at lower energies at ISR[25-29] are shown in Figs. 1(a) and 1(b).

The PHENIX and STAR spectra are compatible each other over an overlapping range. The p_T-spectra of π^0-meson production demonstrate strong dependence on the collision energy. As seen from Fig. 1(b) the new data on inclusive cross sections of π^0-mesons obtained at RHIC are in a good agreement with our earlier results[17] .

The shape of the scaling function $\psi(z)$ for ISR and RHIC energies is the same. The uncertainty of the relative normalization factor for the cross sections is about 2.

Based on the obtained results we conclude that the available experimental data on high-p_T π^0-meson production in pp collisions confirm the energy independence of $\psi(z)$ in a large region of z.

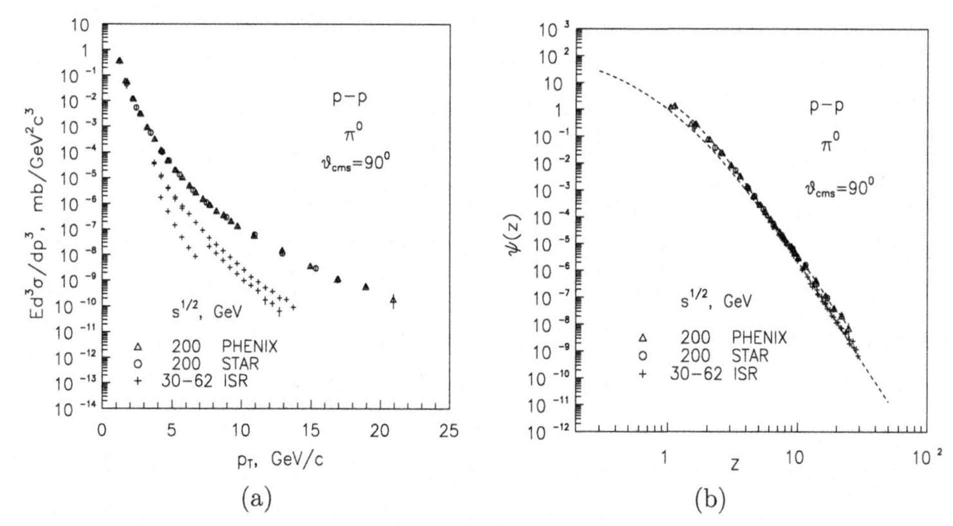

Fig. 1. Spectra of π^0 mesons produced in pp collisions in p_T and z-presentations. Experimental data are taken from[23,24] .

3.2. *Charged hadrons*

Here we present results of the joint analysis of experimental data on charged hadrons produced in $\bar{p}p$[33] and pp[34] collisions at different multiplicities of charged particles and different incident energies over a wide p_T range.

The E735 collaboration measured the multiplicity dependence of charged hadron spectra[33] in proton–antiproton collisions at the energy $\sqrt{s} = 1800$ GeV for $dN_{\rm ch}/d\eta = 2.3 - 26.2$ at Tevatron. The measurements include highest multiplicity density per nucleon–(anti)nucleon collision obtained so far. The pseudorapidity range was $|\eta| < 3.25$. The data cover the transverse momentum range $p_T = 0.15 - 3$ GeV/c. The spectra demonstrate strong sensitivity on multiplicity density at high p_T. The independence of the function $\psi(z)$ on the multiplicity density $dN_{\rm ch}/d\eta$ was obtained. The result gives restriction on the parameter c which value was found to be $c \simeq 0.25$.

The STAR collaboration obtained new data[34] on the inclusive cross sections of charged hadrons produced in proton-proton collisions in the central rapidity range $|\eta| < 0.5$ at the energy $\sqrt{s} = 200$ GeV at RHIC. The transverse momentum spectra were measured up to $p_T = 9.5$ GeV/c. The spectra demonstrate strong dependence on the multiplicity density for $dN_{\rm ch}/d\eta = 2.5, 6.0$ and 8.0. The STAR data confirm the multiplicity independence of the scaling function $\psi(z)$ established for the proton-antiproton collisions at higher energies. The result was obtained with the same value of the parameter $c = 0.25$.

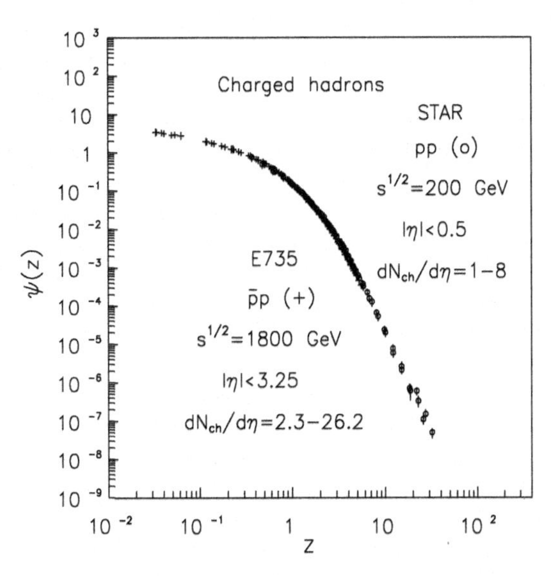

Fig. 2. Spectra of charged hadrons produced in $\bar{p}p$ and pp collisions in z-presentation. Experimental data are taken from[33,34] .

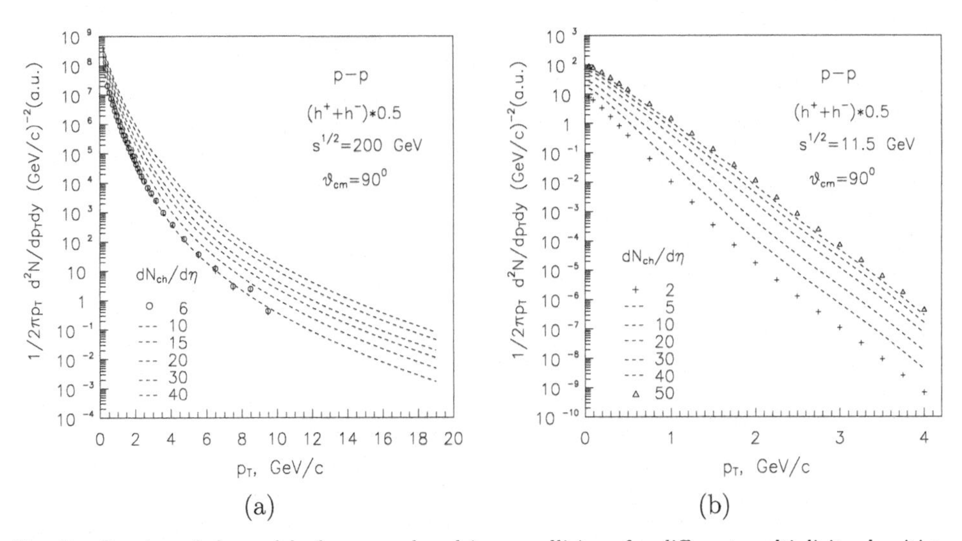

Fig. 3. Spectra of charged hadrons produced in pp collisions for different multiplicity densities $dN_{ch}/d\eta$ in p_T-presentation at RHIC (200 GeV) (a) and U70 (11.5 GeV) (b) energies. Experimental data are taken from[34] .

Similarly as E735 and STAR data, the experimental data on multiplicity dependence of the spectra for charged hadrons obtained by UA1[31] and CDF[35] collaborations at the S$\bar{\text{p}}$pS and Tevatron are found to be in good agrement with the scaling function $\psi(z)$ for $c = 0.25$. The scaling for charged particles produced in proton–antiproton and proton–proton collisions for different multiplicities and energies is consistent with the values of the fractal dimensions $\delta_1 = \delta_2 = 0.7$ and $\epsilon = 0.7$.

Figure 2 illustrates the results of the combine analysis of E735 and STAR data on the inclusive cross sections in the z-presentation. The STAR data demonstrates power law in the region of high z. The behavior of $\psi(z)$ for $\bar{p}p$ and pp collisions coincides each other in the overlapping range of $z = 0.4 - 5$. There is indication that in the low z-range the shape of the scaling function is the same for both collisions. The spectra for $\bar{p}p$ and pp collisions are used for construction of the scaling function $\psi(z)$ over a wide range of z.

We use the properties of the z-scaling to predict the cross sections of charged hadron production in pp collisions over a wide range of the transverse momenta, multiplicity densities, and collision energies. Figure 3 shows multiplicity dependence of the spectra of charged particles produced in pp collisions in the central rapidity range at $\sqrt{s} = 200$ and 11.5 GeV. The predictions are of interest for searching for phase transitions of hadron matter at extremely high multiplicity (energy) densities and can be verified at U70 and RHIC. We suppose that violation of the z-scaling in particle production can be signature of a phase transition.

3.3. *Mean p_T vs. multiplicity and collision energy*

The correlation between $< p_T >$ and multiplicity density of charged particles $dN_{\text{ch}}/d\eta$ in high-energy hadron collisions was experimentally observed at ISR[30] , S$\bar{\text{p}}$pS[31] , and Tevatron[32,33,35] . The dependence of $< p_T >$ on the collision energy $\sqrt{s}$ was experimentally established as well. It was found that $< p_T >$ grows with the multiplicity, collision energy, and mass of inclusive particles.

The average transverse momentum of the produced particles characterizes medium created in the collisions of hadrons or nuclei. The dependence of $< p_T >$ on the collision energy $\sqrt{s}$ and multiplicity density $dN_{\text{ch}}/d\eta$ is useful tool to investigate the collective behavior of soft multi-particle production. It is believed that this could give indications on phase transitions in hadron matter at extremely high multiplicity densities. On the other hand, as noted in Ref.[35] , different theoretical models used for explanation of the observed phenomena connected with mechanism of multi-particle production do not provide satisfactory predictions for existing experimental results leaving real origin of the effects unexplained. Therefore development of new methods of data analysis to clarify features of multi-particle production is of interest.

We propose to explore the energy and multiplicity independence of the scaling function $\psi(z)$ for charged hadrons to study the dependence of $< p_T >$ on $dN_{\text{ch}}/d\eta$ and $\sqrt{s}$. The results obtained in previous section allow us to construct the transverse momentum distributions dN/dp_T over a wide kinematical range and calculate the mean p_T as follows

$$< p_T >= \frac{\int_{p_T^{\min}}^{\infty} p_T(dN/dp_T)dp_T}{\int_{p_T^{\min}}^{\infty}(dN/dp_T)dp_T}. \tag{9}$$

Here p_T^{min} is the minimal transverse momentum of the detected particle. Figure 4 demonstrates dependence of $< p_T >$ for the produced charged hadrons on the multiplicity density $dN_{\rm ch}/d\eta$ and the collision energy $\sqrt{s}$. The multiplicity and energy dependencies of $< p_T >$ reveal monotonous growth with $dN_{\rm ch}/d\eta$. Similar behavior of $< p_T >$ were found for the strange particles (K_S^0, Λ) produced in pp collisions at RHIC energies[36] . We observe no indications of a jump or a sharp rise of $< p_T >$ up to the highest values of the multiplicity density. Experimental verification of these predictions is of interest in search for phase transitions of nuclear matter.

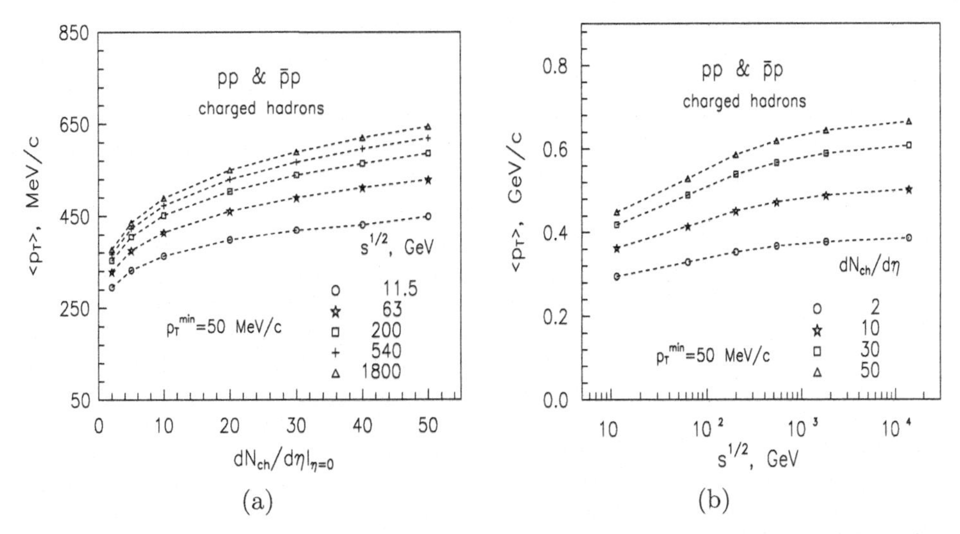

Fig. 4. Mean transverse momentum $< p_T >$ of the charged particles as a function of the multiplicity density $dN_{ch}/d\eta$ (a) and the collision energy $\sqrt{s}$ (b).

3.4. *Jets*

The STAR collaboration published new data[37] on cross sections of jet production in pp collisions at RHIC. The data cover the kinematical range of the pseudorapidity $0.2 < \eta < 0.8$ and the transverse momentum $p_T = 5 - 50$ GeV/c. Figure 5(a) shows the inclusive differential cross section for the $p + p \to jet + X$ process at $\sqrt{s} = 200$ GeV measured by STAR collaboration. The NLO QCD calculated results with the CTEQ6M parton distribution functions at equal factorization and renormalization scales $\mu_R = \mu_F = p_T$ demonstrate satisfactory agreement with the data. A comparison of the STAR data with MC results[38] and predictions of the

z-scaling is shown in Fig. 5(b). Sensitivity of the scaling function to the choice of the parameters E_{seed}, R, E_{cut} of the jet-finding algorithm is observed. The sensitivity enhances as the transverse momentum of jet decreases. The dependence of η^{jet} on p_T^{jet} was used for construction of the scaling function. Note that the shape of the scaling function $\psi(z)$ for $p_T < 10$ GeV/c can not be described by the power law $\psi(z) \sim z^{-\beta}$. Both MC simulation results and STAR data are in a good agreement with the z-scaling predictions for $p_T^{Jet} > 25$ GeV/c ($z > 180$). The value of the slope parameter β is found to be $\beta = 6.01 \pm 0.06$ for Monte Carlo results at $\sqrt{s} = 200$ GeV. It is compatible with $\beta = 5.95 \pm 0.21$ obtained at lower energies $\sqrt{s} = 38.8, 45, 63$ GeV[19] .

Measurements of jet spectra in *pp* collisions at high p_T with higher accuracy are necessary for precise test of QCD and for verification of the asymptotic behavior of $\psi(z)$ predicted by the z-scaling.

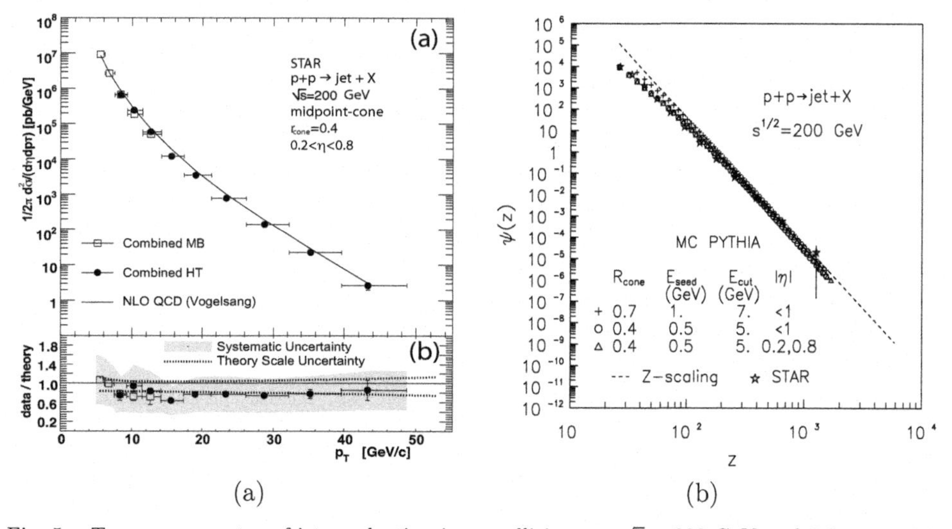

(a) (b)

Fig. 5. Transverse spectra of jet production in *pp* collisions at $\sqrt{s} = 200$ GeV and $0.2 < \eta < 0.8$ in p_T (a) and z (b) presentation. Experimental data obtained by STAR collaboration are taken from[37] . MC results and z-scaling prediction are shown by points ($+, \circ, \triangle$) and the dashed line, respectively.

3.5. *Similarity of charged hadron production in AA*

The multiplicity dependence of p_T and z-presentations of the experimental data[39,40] on inclusive cross sections of charged hadrons produced in heavy ion collisions at RHIC are of special interest.

The multiplicity density $dN/d\eta$ is important ingredient of the z-scaling. It is function of the collision energy $\sqrt{s}$ and the pseudorapidity η. Definition of the scaling function $\psi(z)$ includes multiplicity density $dN/d\eta$ of the corresponding particle species as function of the pseudorapidity η. The scaling variable z is defined in terms of the charged hadron multiplicity density $dN_{ch}/d\eta|_0$ at $\eta = 0$. The multiplicity density $dN_{ch}/d\eta|_0$ characterizes special selection of events. The transverse momentum spectra of hadrons were measured by the E735 collaboration up to the highest $dN_{ch}/d\eta|_0 \simeq 26$[33] . The strong sensitivity of the spectra to $dN_{ch}/d\eta|_0$ was observed with the increasing p_T. The difference between cross sections corresponding to the highest and lowest multiplicity density was found to be about one order of magnitude. The obtained value of the multiplicity density of selected events for $\bar{p}p$ collisions was larger than $dN_{ch}/d\eta|_0/(0.5N_p)$ measured in the central nucleus-nucleus collisions at AGS, S$\bar{\text{p}}$pS, and RHIC. The multiplicity density in the central $PbPb$ collisions at LHC energies is expected to be about 8000 particles per one unit of pseudorapidity. Such regime of particle production at very high multiplicity densities is believed to be preferable for detailed studies of QGP properties.

Figure 6 shows dependence of the spectra of the charged hadron production in $AuAu$ collisions on the transverse momentum p_T at the energy $\sqrt{s} = 200$ GeV over the pseudorapidity range $|\eta| < 0.5$ for different centralities[39] . The data cover a wide transverse momentum range $p_T = 0.2 - 11$ GeV/c. Strong sensitivity of the high-p_T spectra to the multiplicity density $dN_{ch}/d\eta|_0$ is observed. We studied the dependence of the transverse momentum spectra on multiplicity density $dN_{ch}/d\eta|_0$ in the framework of data z-presentation (Fig.7(a)). We have used the values of the fractal dimensions $\delta_N = 0.7$, $\delta_A = A\delta_N$, and $\epsilon = 0.7$. The scaling behavior of the data is restored for $c = 0.25$ with simultaneous utilization of the transformation: $z \to \alpha_\rho z$ and $\psi \to \alpha_\rho^{-1}\psi$ $(\rho \equiv dN_{ch}/d\eta|_0)$. The power behavior (the straight dashed line in Fig.7(a)) of the scaling function, $\psi(z) \sim z^{-\beta}$, for high z is observed. The soft regime of particle production demonstrates self-similarity in z-presentation for low z as well.

Figure 7(b) demonstrates z-presentation of the spectra of charged hadrons produced in $CuCu$ collisions in the central rapidity range at $\sqrt{s} = 200$ GeV as a function of multiplicity density. The data obtained by the PHOBOS collaboration[40] reach the transverse momentum $p_T = 4$ GeV/c. The scaling was found to be restored with the same value of $c = 0.25$. The dashed lines are fits of the function $\psi(z)$ corresponding to the STAR data[39] at $\sqrt{s} = 200$ GeV. We have used the transformation $z \to \alpha_A z$ and $\psi \to \alpha_A^{-1}\psi$ for comparison of the scaling functions for different nuclei. The z-presentation demonstrates good compatibility of the PHOBOS and STAR data over the kinematical range measured by the PHOBOS collaboration[40] .

Based on the obtained results we conclude that the mechanism of charged hadron production in $AuAu$ and $CuCu$ collisions at $\sqrt{s} = 200$ GeV reveals self-similarity and fractality at the level of cross sections over a wide range of the transverse momentum and multiplicity density.

4. z-Scaling in $\bar{p}p$ collisions at Tevatron

In this section we present results of our analysis of new data obtained by the D0 and CDF Collaborations at Tevatron in Run II. We verify properties of z-scaling

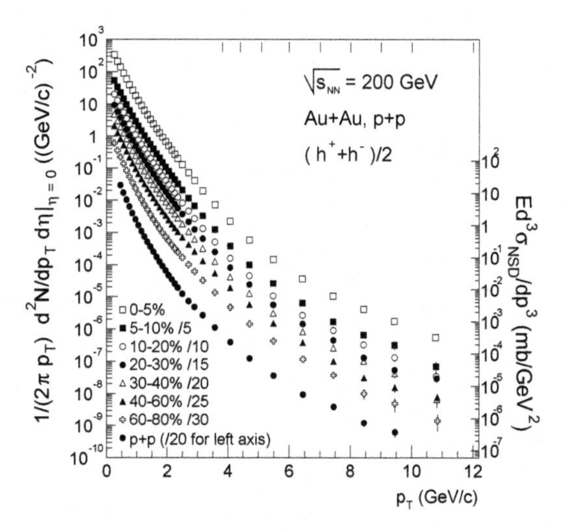

Fig. 6. Transverse momentum spectra of charged hadrons produced in $AuAu$ and pp collisions at RHIC as a function of multiplicity density. Experimental data obtained by STAR Collaboration are taken from[39] .

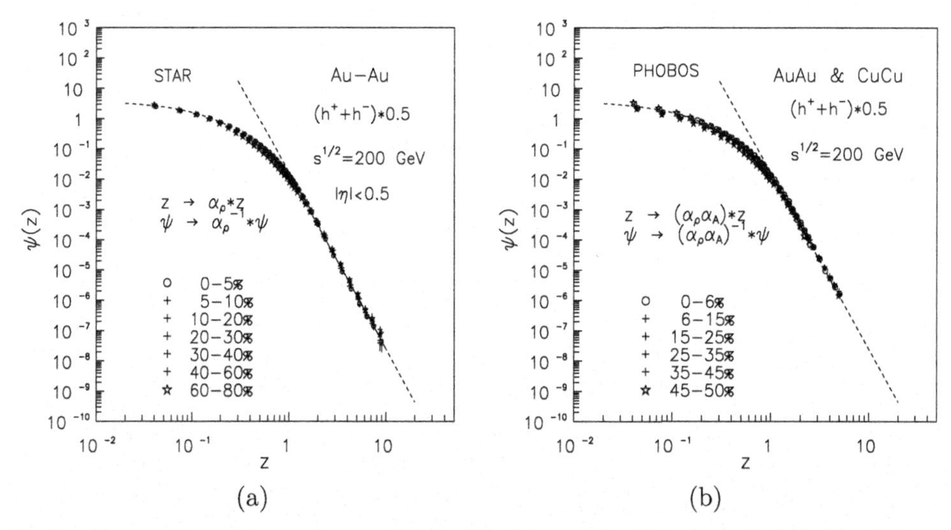

Fig. 7. Transverse momentum spectra of charged hadrons produced in $AuAu$ and $CuCu$ collisions at RHIC in z-presentation as a function of multiplicity density. Experimental data are taken from[39,40] .

established in previous papers such as the energy and angular independence of the scaling function $\psi(z)$ for particle (hadrons, direct photons, jets) production at high p_T. The hypothesis of flavor independence of $\psi(z)$ for high z is used for prediction of spectra of different particles ($J/\psi, \Upsilon, D^0, B^+, Z, W^+$) as well.

4.1. *Direct photons*

Recently the D0 collaboration published the new data[41] on inclusive cross sections of direct photons produced in $\bar{p}p$ collisions at $\sqrt{s} = 1960$ GeV. The data cover the momentum $p_T = 30 - 250$ GeV/c and pseudorapidity $|\eta| < 0.9$ range. These data together with the data obtained by D0 collaboration in Run I are presented in Fig. 8(a). A strong angular dependence of the cross section is observed. It increases with p_T and reaches about one order of magnitude at $p_T = 100$ GeV/c.

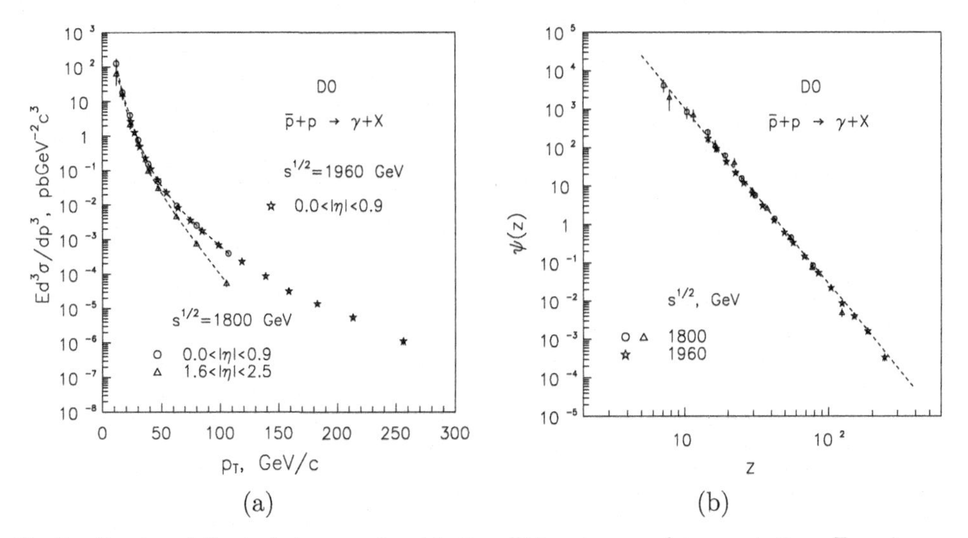

Fig. 8. Spectra of direct photons produced in $\bar{p}p$ collisions in p_T and z-presentations. Experimental data obtained by D0 Collaboration in Run I and II are taken from[41] .

The z-presentation of the same data is shown in Fig. 8(b). One can see that new experimental data confirm the features (energy and angular independence of $\psi(z)$) of the z-scaling for direct photon production in $\bar{p}p$ and pp collisions established in[18,42] . The power law, $\psi(z) \sim z^{-\beta}$, is observed over a wide range of z. We consider the independence of the slope parameter β on kinematical variables ($\sqrt{s}, p_T$ and η) as an evidence of self-similarity and fractality of photon production. It can mean that structure of a photon itself at small scales looks like structure of other particles (hadrons) characterized by fractal dimension(s) †.

† In paper[18] the fractal dimensions δ_1, δ_2 are introduced as parameters for description of the fractal measure $z = z_0\Omega^{-1}$, $\Omega = (1 - x_1)^{\delta_1}(1 - x_2)^{\delta_2}$

4.2. *Jets*

In this section we present results of analysis of new data on inclusive cross sections of jet production in $\bar{p}p$ collisions at $\sqrt{s} = 1960$ GeV obtained by the D0 and CDF collaborations at Tevatron[43] and compare them with our previous results[19] .

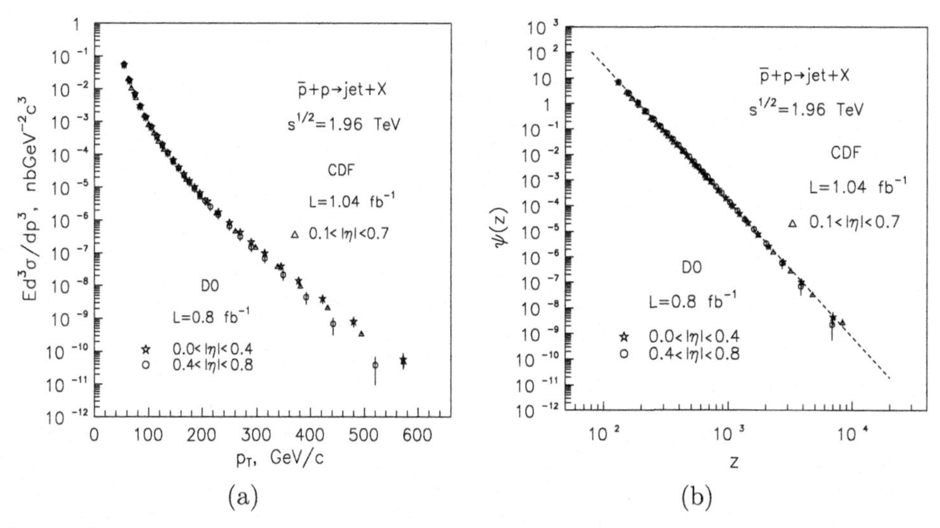

Fig. 9. Spectra of jet production in $\bar{p}p$ collisions in p_T and z-presentations. Experimental data obtained by the D0 and CDF collaborations are taken from[43] .

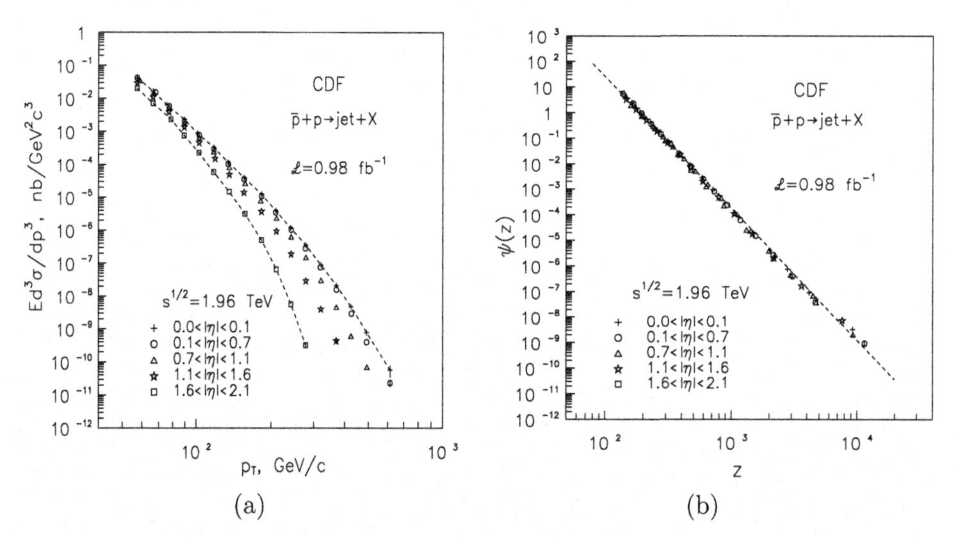

Fig. 10. Angular dependence of spectra of jet production in $\bar{p}p$ collisions in p_T and z-presentations. Experimental data are taken from[43] .

Production of hadron jets at Tevatron probes the highest momentum transfer region currently accessible which is potentially sensitive to a wide variety of new physics. The information on inclusive jet cross sections at high transverse momentum range forms a basis to test QCD, to extract the parton distribution functions and to constrain uncertainties for gluon distribution in the high-x range. In Run II, as mention in[44] , the measurement of jet production and the sensitivity to new physics will profit from the large integrated luminosity and the higher cross section, which is associated with the increase in the center-of-mass energy from 1800 to 1960 GeV. Therefore a test of the z-scaling for jet production in $\bar{p}p$ collisions in new kinematic range is of great interest to verify scaling features established in our previous analysis[19] .

Figure 9(a) shows the new Run II data[43] on the inclusive jet cross section at $\sqrt{s} = 1960$ GeV. The pseudorapidity range covered by the D0 and CDF collaborations corresponds to $|\eta| < 0.8$ and $0.1 < |\eta| < 0.7$, respectively. The transverse momentum of jet changes from 50 up to 560 GeV. The z-presentation of data is shown in Fig.9(b). As seen from Fig.9(b) the D0 and CDF data are compatible each other. The energy independence of $\psi(z)$ is observed up to $z \simeq 1000$. Asymptotic behavior of the scaling function is described by the power law, $\psi(z) \sim z^{-\beta}$ (the dashed line in Fig.9(b)). The slope parameter β is energy independent over a wide p_T-range. Note that results of present analysis of new D0 data are in a good agreement with our results[19] based on the data obtained by the same collaboration in Run I. The energy independence and the power law (the dashed line in Fig.9(b)) of the scaling function $\psi(z)$ are found to be as well.

The angular dependence of inclusive cross section of jet production in $\bar{p}p$ collisions at $\sqrt{s} = 1960$ GeV was investigated by the CDF collaboration[43] . The experimental data cover the rapidity range $|\eta| < 2.1$. The highest transverse energy carried by one jet was determined to be 600 GeV. As seen from Fig. 10(a) the transverse momentum spectra demonstrate strong dependence on the pseudorapidity of the produced jet. The z-presentation of the same data is shown in Fig. 10(b). It demonstrates the angular independence and the power behavior of $\psi(z)$. We would like to emphasize that these results are new confirmation of the z-scaling. Jet production is usually considered as a signature of hard collisions of the elementary hadron constituents (quarks and gluons). Therefore the obtained result means that the interaction of the constituents, their substructure and mechanism of jet formation reveal the properties of self-similarity over a wide scale range (up to 10^{-4} Fm).

4.3. *b-Jets*

The flavor independence of the z-scaling in the large z region means that value of the slope parameter β of the scaling function $\psi(z)$ is the same for different types of the produced hadrons. The hypothesis is supported by results of analysis of hadron $(\pi^{\pm,0}, K, \bar{p})$ spectra for high p_T in pp and pA collisions[45] . The verification of the hypothesis is of interest for understanding of mechanism of particle production at very

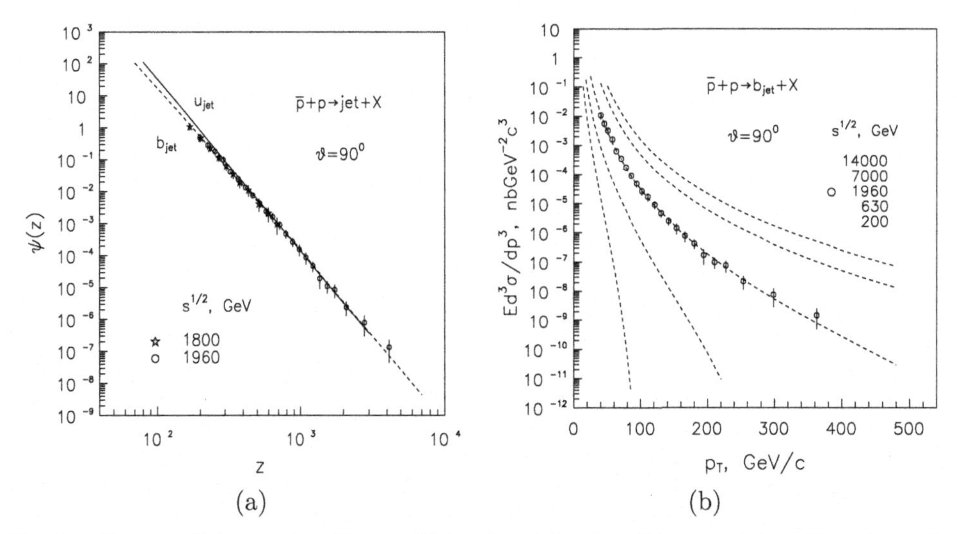

Fig. 11. Spectra of b-jets produced in $\bar{p}p$ collisions in z (a) and p_T (b) presentations. Experimental data are taken from[46] .

small scale. We assume that the transformation of a point-like quark (u, d, s, c, b, t) into real hadron produced at high p_T is a self-similar process which is independent of the quark flavor.

The spectra of b-tagged jets were measured[46] by the D0 and CDF collaborations at $\sqrt{s} = 1800$ and 1960 GeV, respectively. Transverse energy of the jet was measured in the range $25 - 400$ GeV. These data together with the CDF data on inclusive cross section for the non-tagged (u_{jet}) jets obtained in Run I are shown in z-presentation in Fig. 11(a). Both data demonstrate power behavior of $\psi(z)$ for high z. Deviation from the power law is observed for $z < 300$ ‡. The present calculation was performed § with the values of the fractal dimensions $\delta_1 = \delta_2 = 1$. We use the universality of asymptotic behavior of $\psi(z)$ for non-tagged jets and pre-asymptotic behavior of $\psi(z)$ for b-tagged jets for construction of the scaling function over a wide range of z. Our predictions of inclusive cross sections of b-jet production in $\bar{p}p$ collisions at $\sqrt{s} = 200 - 14000$ GeV and the CDF data (o) are shown in Fig. 11(b).

4.4. J/ψ mesons

Here we present results of analysis[47] of the data[48,49] on inclusive cross section of J/ψ production in $\bar{p}p$ collisions in the region of large transverse momenta. The hypothesis of the flavor independence of z-scaling was used for the analysis. The experimental UA1 data[50] on inclusive cross section of π^0-mesons produced in $\bar{p}p$ collisions at $\sqrt{s} = 540$ GeV were used to construct the scaling function $\psi(z)$ in the asymptotic

‡We assume that for s- and c-tagged jets the deviation will be smaller and for t-tagged jets larger than for b-tagged jets.
§The calculation procedure is described in[19] .

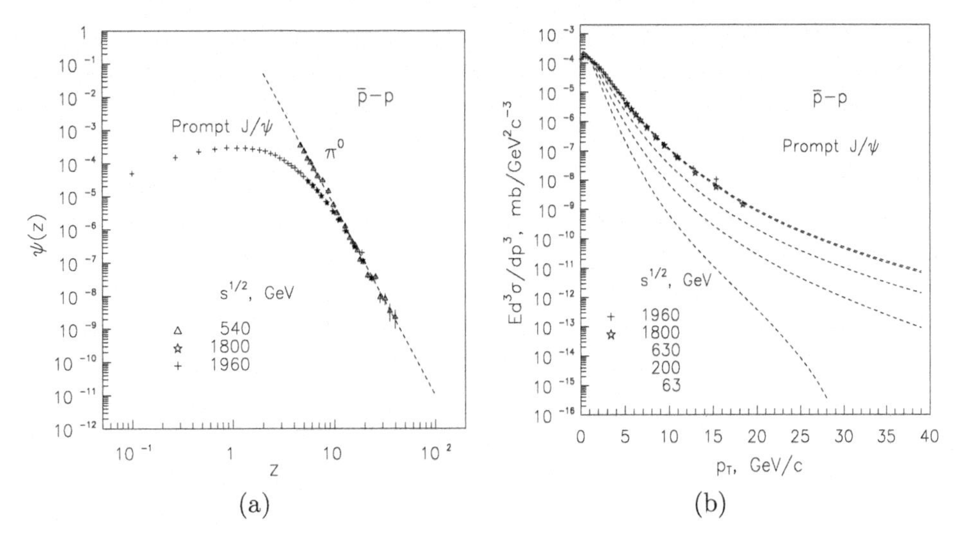

(a) (b)

Fig. 12. Spectra of J/ψ mesons produced in $\bar{p}p$ collisions in z (a) and p_T (b) presentations. Experimental data are taken from[48–50] .

region. The scaling function of π^0 is shown by the dashed line in Fig.12(a). It is described by the power law, $\psi(z) \sim z^{-\beta}$. The value of the slope parameter β was found to be 5.77 ± 0.02 over a wide p_T range. To compare z-presentations for J/ψ and π^0 in the large z region, the transformation $z \to \alpha_F \cdot z, \quad \psi \to {\alpha_F}^{-1} \cdot \psi$ was applied. The coefficient α_F is independent on kinematical variables. Its value was obtained to describe the overlapping region for both particles. The normalization of $\psi(z)$ is here unessential. The energy independence of the scaling function was used to predict transverse spectra of J/ψ production in $\bar{p}p$ collisions in the central rapidity range at the energy $\sqrt{s} = 63, 200, 630, 1800$, and 1960 GeV. The calculated results and the CDF data[49] are shown in Fig. 12(b) by the dashed lines and points, respectively. One can see that strong dependence of the inclusive cross section on $\sqrt{s}$ enhances with p_T. Experimental test of the predicted results is of interest for understanding of mechanism of J/ψ production and for verifying of properties of the z-scaling.

4.5. D^0 *mesons*

Data on open charm production can provide understanding of mechanism of particle formation depending on the flavor quantum numbers and test of QCD predictions. They could give additional constraints on parton distribution and fragmentation functions of the charmed quark. Such data allow us to study flavor dependent features of z-scaling.

The inclusive charm meson cross sections were measured in $\bar{p}p$ collisions by the CDF collaboration in the central rapidity $|y| < 1$ and transverse momentum $p_T = 5-20$ GeV/c range at $\sqrt{s} = 1960$ GeV[51] . Here we present results of data analysis corresponding to the reconstructed decay mode $D^0 \to K^-\pi^+$ in the framework of z-scaling. It was established that the prompt fraction of D^0 meson production for each p_T bin is $(86.6 \pm 0.4)\%$. The hypothesis of the flavor and energy independence

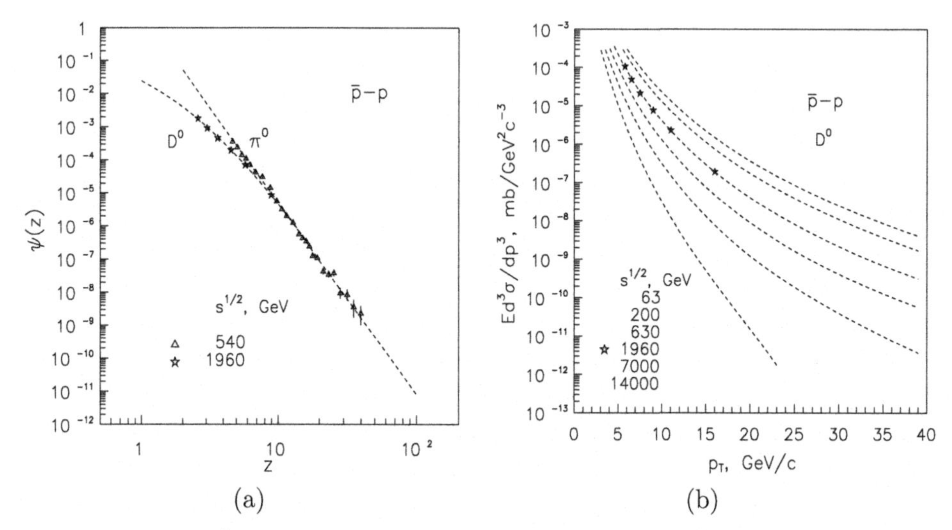

Fig. 13. Spectra of D^0 mesons produced in $\bar{p}p$ collisions in z and p_T presentations. Experimental data are taken from[50,51] .

of z-scaling for D^0 meson production was used in the analysis. The data[50] on inclusive cross section of π^0-mesons produced in $\bar{p}p$ collisions at $\sqrt{s} = 540$ GeV and the CDF data[51] were used for construction of the scaling function for D^0 mesons over a wide range of z. The results are shown in Fig. 13(a). The dashed lines are obtained by fitting the data[50,51] in z-presentation. Figure 13(b) shows transverse momentum spectra predicted by the z-scaling for the prompt D^0 meson production in $\bar{p}p$ collisions over the kinematical range $\sqrt{s} = 63 - 14000$ GeV, $p_T = 5 - 40$ GeV/c, and $\theta_{cms} = 90^0$. The obtained results are of interest for comparison with QCD predictions and with experimental data over a wider range of p_T.

4.6. B^+ *mesons*

The first direct measurements of B meson differential cross sections in $\bar{p}p$ collisions at $\sqrt{s} = 1800$ GeV were presented in[52] . The identification was performed by measuring the mass and momentum of the B meson decaying into exclusive final states. The cross section was measured in the central rapidity region $|y| < 1$ for $p_T(B) > 6.0$ GeV/c. We have analyzed the data on transverse momentum spectrum of B^+ mesons reconstructed via the decay $B^+ \to J/\psi K^+$ with $J/\psi \to \mu^+\mu^-$ in the framework of z-scaling.

Figure 14(a) shows the experimental data on inclusive spectra for B^{+}[52] and π^{+}[50] mesons produced in $\bar{p}p$ collisions in z-presentation. The normalization of $\psi(z)$ is here unessential. The predictions of inclusive cross sections for B^+ mesons over

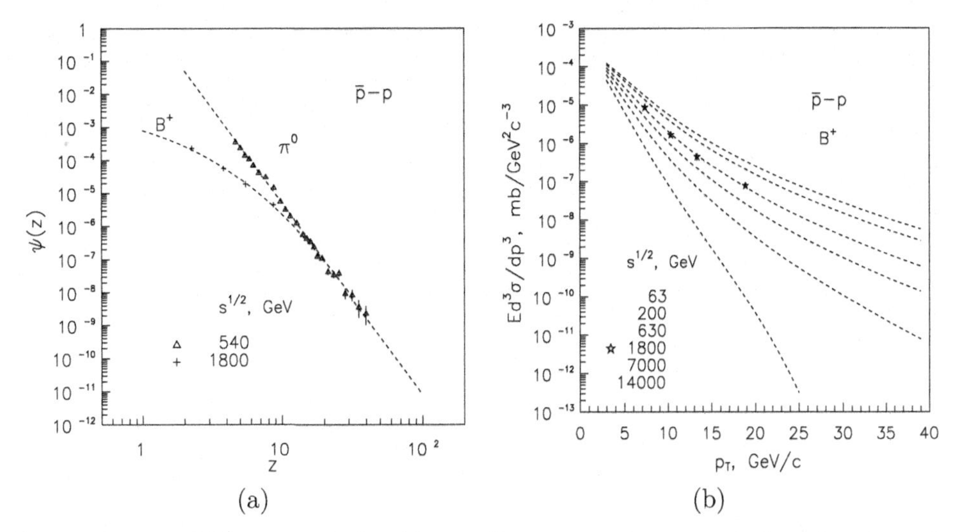

Fig. 14. Spectra of B^+ mesons produced in $\bar{p}p$ collisions in z and p_T presentations. Experimental data are taken from[50,52] .

a range $\sqrt{s} = 63 - 14000$ GeV, $\theta_{cms} = 90^0$, and $p_T = 5 - 40$ GeV/c are plotted in Fig. 14(b). As seen from Fig. 14 data on transverse momentum spectrum for $p_T > 25$ GeV/c is necessary to determine the overlapping range and study the asymptotic behavior of $\psi(z)$.

4.7. $\Upsilon(1S)$ *mesons*

Features of heavy flavor production in high energy hadron collisions at high p_T can be related to new physics phenomena at small scales. Understanding of flavor origin and search for similarity of particle properties depending on the additive quantum numbers (strangeness, charm, beauty, top) is fundamental problem of particle physics.

The differential cross sections of Υ production in $\bar{p}p$ collisions in the rapidity range $|y| < 0.4$ in $1S, 2S$, and $3S$ states at $\sqrt{s} = 1800$ GeV are presented in[53] . All three resonances were reconstructed through the decay $\Upsilon \to \mu^+\mu^-$. Transverse momentum of Υ was measured over the range $p_T = 0.5 - 20$ GeV/c. The shape of p_T spectrum was found to be the same for all states. The data are important for the investigation of the bound state production mechanisms in $\bar{p}p$ collisions.

Figure 15 demonstrates results of our analysis of experimental data on $\Upsilon(1S)$ production in $\bar{p}p$ collisions. The function $\psi(z)$ for $\Upsilon(1S)$ (see Fig.15(a)) was constructed using both the $\Upsilon(1S)$ [53] and π^0 [50] data on transverse momentum spectra. Our predictions of the inclusive cross sections over a range $\sqrt{s} = 63 - 14000$ GeV, $\theta_{cms} = 90^0$, and $p_T = 1 - 40$ GeV/c are shown in Fig. 15(b). Comparison of the obtained spectra with experimental data and QCD calculations are of interest for verification of the flavor independence of the z-scaling and test of mechanisms of vector meson production in the framework of the QCD theory.

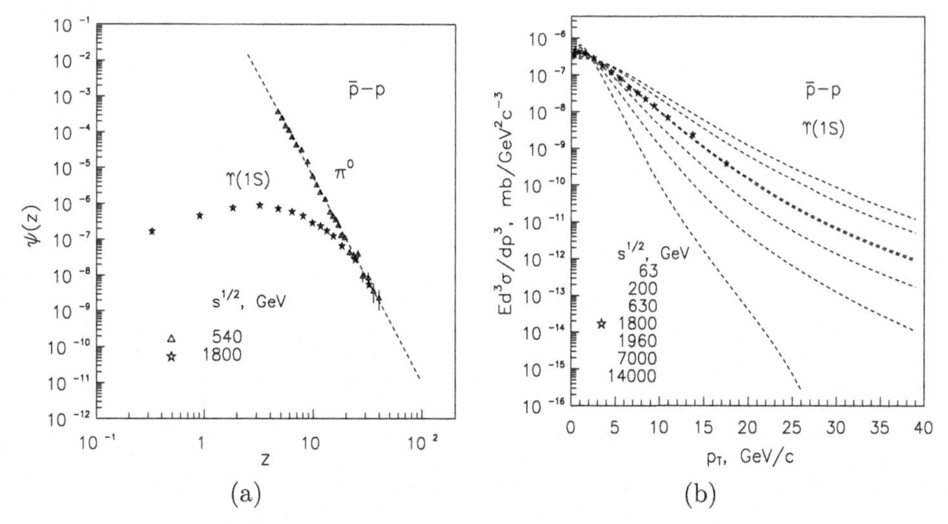

Fig. 15. Spectra of $\Upsilon(1S)$ mesons produced in $\bar{p}p$ collisions in z and p_T presentations. Experimental data are taken from[50,53] .

4.8. Z^0, W^+ *bosons*

Vector Z and W bosons are carriers of the electroweak and strong interactions. High precision data on the transverse momentum spectra for both Z and W bosons produced in high energy $\bar{p}p$ collisions can provide important tests and direct confirmation of the unified model of the weak, electromagnetic, and strong interactions (the Standard Model). Quantum chromodynamics ascribes the transverse momentum of the vector bosons to the associated production of one or more gluons or quarks with the boson. The large mass of the vector bosons assures a large energy scale for probing perturbative QCD with good reliability. It also provides bounds on parametrizations of the parton distribution functions used to describe the non-perturbative regime of the QCD processes. Deviations from the prediction for high p_T could indicate new physics phenomena beyond the Standard Model. Measurements of the differential cross sections for Z and W boson production as a function of transverse momentum in $\bar{p}p$ collisions at $\sqrt{s} = 1800$ GeV over a range $p_T = 1 - 200$ GeV/c are presented in[54,55] . The W and Z bosons were detected through their leptonic decay modes ($W \to e\nu, Z \to e^+e^-$).

We assume that asymptotic behavior of the scaling function for the vector bosons and direct photons produced in $\bar{p}p$ collisions is similar. We consider that this can be described by the power law for high z with the same value of the slope parameter β of $\psi(z)$ for both particles. The similarity of direct and virtual photon (Drell-Yan pair) with vector boson production for high p_T is considered as an important feature of constituent interactions. The asymptotic behavior of $\psi(z)$ of direct photons was used to construct $\psi(z)$ for vector bosons in the range of large z. Figures 16(a) and 17(a) demonstrate z-presentation of data on inclusive transverse momentum spectra for vector bosons[54,55] and direct photons[18] . Note that the overlapping range of $\psi(z)$

for the W^+ boson and γ is large enough. Nevertheless direct measurements of the transverse spectra for W^+ bosons for $p_T > 200$ GeV/c are necessary for verification of this assumption. Figures 16(b) and 17(b) show predictions of the inclusive cross sections of Z and W^+ bosons in $\bar{p}p$ collisions over a range $\sqrt{s} = 200 - 14000$ GeV, $p_T = 1 - 500$ GeV/c and $\theta_{cms} = 90^0$.

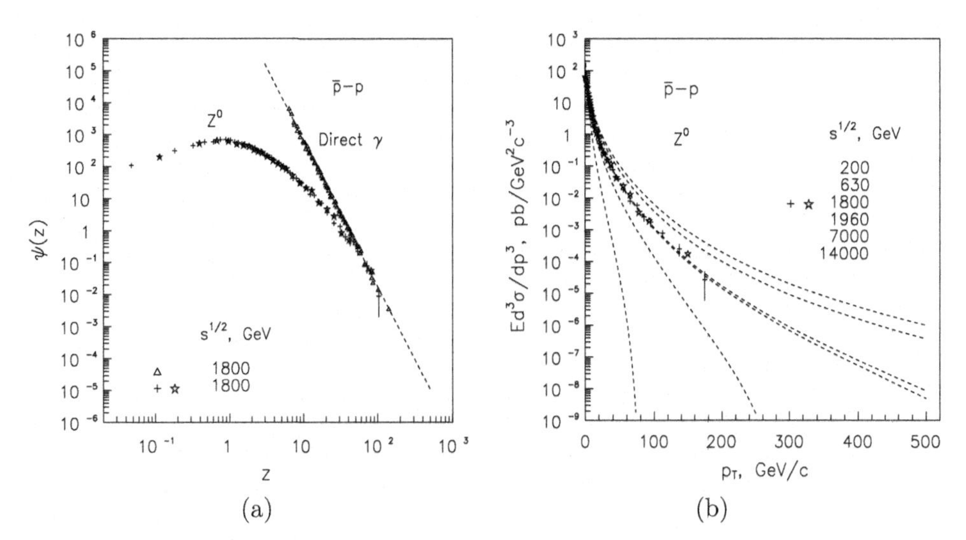

Fig. 16. Spectra of Z^0 bosons produced in $\bar{p}p$ collisions in z and p_T presentations. Experimental data are taken from[42,54] .

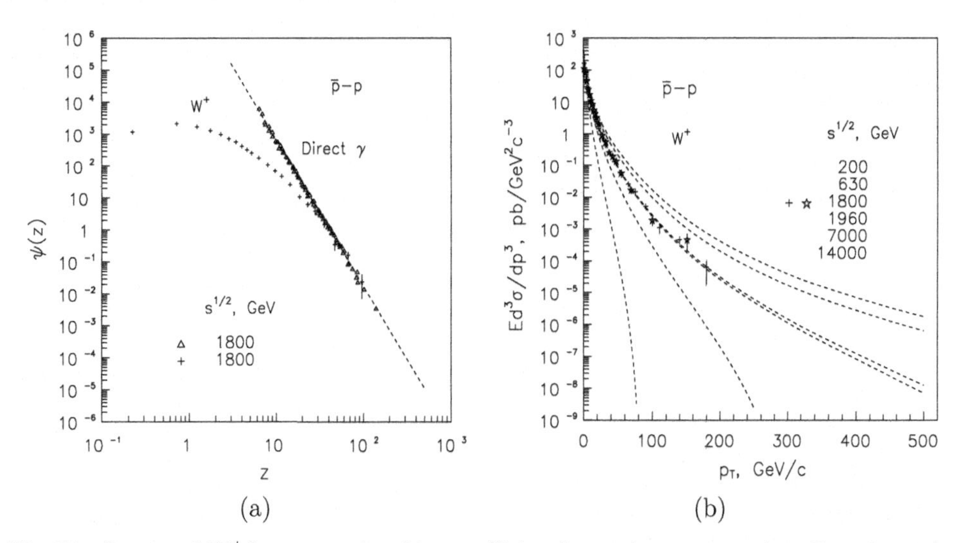

Fig. 17. Spectra of W^+ bosons produced in $\bar{p}p$ collisions in z and p_T presentations. Experimental data are taken from[42,55] .

5. Conclusions

The results of recent analysis of experimental data on inclusive cross sections of particle production in pp, AA, and $\bar{p}p$ collisions at high p_T at RHIC and Tevatron were reviewed. The analysis was performed in the framework of the z-scaling. Physical concept of the scaling and interpretation of the function $\psi(z)$ and the variable z were discussed. It was shown that the generalized concept of z-scaling allows us to study the multiplicity dependence of particle spectra and restore the multiplicity independence of the scaling function. The properties of z-presentation of experimental data were verified. We consider that the properties established in the present analysis of data reflect general features of the structure of the colliding objects, interaction of their constituents, and mechanism of particle formation. The hypothesis on universality of the asymptotic behavior of the scaling function (for π^0-mesons, direct photons) was used for construction of $\psi(z)$ for $J/\psi, \Upsilon(1S), D^0, B^+, Z, W^+$ particles in the region of large z and for predictions of the transverse momentum spectra over a wide range of $\sqrt{s}$ and p_T. The obtained results support the main idea of the z-scaling based on the property of self-similarity of particle interactions at high p_T.

We consider that the z-scaling may be used as a tool for searching for new physics phenomena beyond the Standard Model in hadron and nucleus collisions at high transverse momentum and high multiplicity at RHIC, Tevatron, and LHC.

Acknowledgments

The investigations have been partially supported by the IRP AVOZ10480505, by the Grant Agency of the Czech Republic under the contract No. 202/07/0079 and by the special program of the Ministry of Science and Education of the Russian Federation, grant RNP.2.1.1.5409.

References

1. E. Eichten, K. Lane, and M. Peskin, *Phys. Rev. Lett.* **50**, 811 (1983).
   E. Eichten, I. Hinchliffe, K. Lane, and C. Quigg, *Rev. Mod. Phys.* **56**, 4 (1984).
2. I. Antoniadis, in *Proceedings of European School of High-Energy Physics*, Beatenberg, Switzerland, 26 August - 8 September, 2001 (Editors: N. Ellis and J. March-Russul) p.301.
3. C. G. Lester, in *Proceedings of Advanced Studies Institute on "Physics at LHC"*, Czech Republic, Prague, July 6-12, 2003 (Editors: M. Finger, A. Janata, and M. Virius) A303.
4. L. Nottale, *Fractal Space-Time and Microphysics* (World Sci., Singapore, 1993);
   B. Mandelbrot, *The Fractal Geometry of Nature* (Freeman, San Francisco, 1982).
5. I. Arsene *et al.*, *Nucl. Phys. A* **757**, 1 (2005);
   B. B. Back *et al.*, *Nucl.Phys. A* **757**, 28 (2005);
   J. Adams *et al.*, *Nucl. Phys. A* **757**, 102 (2005);
   K. Adcox *et al.*, *Nucl. Phys.A* **757**, 184 (2005).
6. R. P. Feynman, *Phys. Rev. Lett.* **23**, 1415 (1969).
7. J. D. Bjorken, *Phys. Rev.* **179**, 1547 (1969);
   J. D. Bjorken, and E. A. Paschos, *Phys. Rev.* **185**, 1975 (1969).

8. P. Bosted *et al.*, *Phys. Rev. Lett.* **49**, 1380 (1972).
9. J. Benecke *et al.*, *Phys. Rev.* **188**, 2159 (1969).
10. A. M. Baldin, *Sov. J. Part. Nucl.* **8**, 429 (1977).
11. V. S. Stavinsky, *Sov. J. Part. Nucl.* **10**, 949 (1979).
12. G. A. Leksin: Report No. ITEF-147, 1976; G.A. Leksin: in *Proceedings of the XVIII International Conference on High Energy Physics*, Tbilisi, Georgia, 1976, edited by N.N. Bogolubov *et al.*, (JINR Report No. D1,2-10400, Tbilisi, 1977), p. A6-3.
13. Z. Koba, H. B. Nielsen and P. Olesen, *Nucl. Phys. B* **40**, 317 (1972).
14. V. A. Matveev, R. M. Muradyan, and A. N. Tavkhelidze, *Part. Nuclei* **2**, 7 (1971); *Lett. Nuovo Cim.* **5**, 907 (1972); *Lett. Nuovo Cim.* **7**, 719 (1973).
15. S. Brodsky, and G. Farrar, *Phys. Rev. Lett.* **31**, 1153 (1973); *Phys. Rev. D* **11**, 1309 (1975).
16. I. Zborovský, Yu. A. Panebratsev, M. V. Tokarev, and G. P. Škoro, *Phys. Rev. D* **54**, 5548 (1996); I. Zborovský, M. V. Tokarev, Yu. A. Panebratsev, and G. P. Škoro, *Phys. Rev. C* **59**, 2227 (1999); M. V. Tokarev, O. V. Rogachevski, and T. G. Dedovich, Preprint No. E2-2000-90, JINR (Dubna, 2000); M. Tokarev, I. Zborovský, Yu. Panebratsev, and G. Skoro, *Int. J. Mod. Phys. A* **16**, 1281 (2001); M. Tokarev, hep-ph/0111202; M. Tokarev and D. Toivonen, hep-ph/0209069; G. P. Skoro, M. V. Tokarev, Yu. A. Panebratsev, and I. Zborovský, hep-ph/0209071; M. Tokarev, *Acta Physica Slovaca* **54**, 321 (2004).
17. M. V. Tokarev, O. V. Rogachevski, and T. G. Dedovich, *J. Phys. G: Nucl. Part. Phys.* **26**, 1671 (2000).
18. M. Tokarev and G. Efimov, hep-ph/0209013;
    M. V. Tokarev, G. L. Efimov, and D. E. Toivonen, *Physics of Atomic Nuclei* **67**, 564 (2004).
19. M. V. Tokarev, and T. G. Dedovich, *Int. J. Mod. Phys. A* **15**, 3495 (2000);
    M. V. Tokarev, and T. G. Dedovich, *Physics of Atomic Nuclei* **68**, 404 (2005).
20. I. Zborovský, and M. V. Tokarev, *Part. and Nucl., Letters* **3**, 68 (2006); hep-ph/0506003.
21. M. V. Tokarev, and I. Zborovský, In: *Proceedings of XXXVI International Symposium on Multiparticle Dynamics (ISMD2006), September 02-08, 2006, Paraty, Rio de Janeiro, Brazil*; http://www.sbf1.sbfisica.org.br/eventos/extras/ismd2006.
22. I. Zborovský, and M. V. Tokarev, JINR Commun. E2-2006-34, Dubna, 2006, 20p.
23. H. Buesching, *"Hot Quark06", May 15-20, 2006, Sardinia, Italy.*
24. D. Relyea, *RHIC & AGS Annual Users Meeting, June 8-9, 2006, BNL, USA.*
25. A. L. S. Angelis *et al.*, *Phys. Lett. B* **79**, 505 (1978).
26. C. Kourkoumelis *et al.*, *Phys. Lett. B* **83**, 257 (1979).
27. C. Kourkoumelis *et al.*, *Z. Phys.* **5**, 95 (1980).
28. D. Lloyd Owen *et al.*, *Phys. Rev. Lett.* **45**, 89 (1980).
29. K. Eggert *et al.*, *Nucl. Phys. B* **B98**, 49 (1975).
30. A. Breakstone *et al. Z. Phys. C* **33**, 333 (1987).
31. G. Arnison *et al.*, *Phys. Lett. B* **118**, 167 (1982).
32. T. Alexopoulos *et al.*, *Phys. Rev. Lett.* **60**, 1622 (1988).
33. T. Alexopoulos *et al.*, *Phys. Lett. B* **336**, 599 (1994).
34. J. E. Gans, *PhD Thesis, Yale University, USA (2004).*
35. D. Acosta *et al.*, *Phys. Rev. D* **65**, 072005 (2002);
    F. Rimondi *et al.*, *Yad. Fiz.* **67**, 128 (2004).
36. R. Witt, *J. Phys. G: Nucl. Part. Phys.* **31**, S863 (2005);
    B. I. Abelev *et al.*, nucl-ex/0607033.
37. M. Miller, hep-ex/0604001.

B. I. Abelev *et al.*, *Phys. Rev. Lett.* **97**, 252001 (2006); hep-ex/0608030.
38. T. G. Dedovich, *PhD Thesis, JINR, Dubna, Russia (2007).*
39. J. Adams *et al.*, *Phys. Rev. Lett.* **91**, 172302 (2003).
40. B. B. Back *et al.*, *Phys. Lett. B* **578**, 297 (2004);
B. Alver *et al.*, *Phys. Rev. Lett.* **96**, 212301 (2006).
E. Wenger, nucl-ex/0511036.
41. A. Abachi *et al.*, *Phys.Rev.Lett.* **77**, 5011 (1996);
B. Abbott *et al.*, *Phys.Rev.Lett.* **84**, 2786 (2000);
V. M. Abazov *et al.*, *Phys. Lett. B* **639**, 151 (2006).
42. A. Abachi *et al.*, *Phys. Rev. Lett.* **77**, 5011 (1996);
F. Abe *et al.*, *Phys. Rev. Lett.* **68**, 2734 (1992);
F. Abe *et al.*, *Phys. Rev. D* **48**, 2998 (1993).
43. M. D′Onofrio (for CDF & D0 collaborations), in *XX Rencontres de Physique de la Vallee D′Aoste, La Tuile, Italy, March 5-11, 2006*; http://www.pi.infn.it/lathuile/2006/Programme.htm.
D. Bandurin, (for CDF & D0 collaborations), 9th Conference on the Intersections of Nuclear and Particle Physics (CIPANP′06), Rio Grande, Puerto Rico, May 30 June 4, 2006; http://cipanp.physics.uiuc.edu/.
44. M. Tönnesmann, *Eur. Phys. J. C* **33**, s422 (2004).
45. M. Tokarev, in: *Proceedings of the International Workshop "Relativistic Nuclear Physics: from Hundreds of MeV to Tev", Varna, Bulgaria, September 10-16, 2001* (Dubna, JINR, E1,2-2001-290, 2001, 300 p.), V.1, p.280-300.
46. B. Abbott *et al.*, *Phys. Rev. Lett.* **85**, 5068 (2000);
M. D′Onofrio (for D0 & CDF collaborations), XXXXth Rencontres de Moriond, "QCD and Hadronic interactions at high energy", 2-19 March, 2005, Italy; http://moriond.in2p3.fr/QCD/2005/
47. M. Tokarev, hep-ph/0405230.
48. F. Abe *et al.*, *Phys. Rev. Lett.* **69**, 3704 (1992);
F. Abe *et al.*, *Phys. Rev. Lett.* **79**, 572 (1997);
F. Abe *et al.*, *Phys. Rev. Lett.* **79**, 578 (1997).
49. Yu. Gotra, *Ph.D. Thesis (JINR, Dubna, Russia, 2004)*;
D. Acosta *et al.*, *Phys. Rev. D* **71**, 032001 (2005);
D. Acosta *et al.*, *Phys. Rev. Lett.* **96**, 202001 (2006).
50. M. Banner *et al.*, *Phys. Lett. B* **115**, 59 (1982).
51. D. Acosta *et al.*, *Phys. Rev. Lett.* **91**, 241804 (2003).
52. F. Abe *et al.*, *Phys. Rev. Lett.* **75**, 1451 (1995);
D. Acosta *et al.*, *Phys. Rev. D* **65**, 052005 (2002).
53. D. Acosta *et al.*, *Phys. Rev. Lett.* **88**, 161802 (2002).
54. T. Affolder *et al.*, *Phys. Rev. Lett.* **84**, 845 (2000);
F. Abe *et al.*, *Phys. Rev. Lett.* **67**, 2937 (1991);
B. Abbott *et al.*, *Phys.Rev. D* **61**, 032004 (1999);
V.M. Abazov *et al.*, *Phys. Lett. B* **B517**, 299 (2001).
55. F. Abe *et al.*, *Phys. Rev. Lett.* **66**, 2951 (1991);
V. M. Abazov *et al.*, *Phys. Lett. B* **517**, 299 (2001);
B. Abbott *et al.*, *Phys. Lett. B* **513**, 292 (2000).

QUARKONIA MEASUREMENTS IN HEAVY-ION COLLISIONS IN CMS

O. L. KODOLOVA (CMS Collaboration)

Moscow State University, Moscow, Russia

The production of quarkonia is one of the most promising signals at the LHC for the study of the production properties of Quark Gluon Plasma. In addition to the J/ψ the extent to which Υ is suppressed should give much insight into the new state of matter. The large muon acceptance and the high precision tracker make the CMS detector ideal for studies of this physics. The estimations of statistics and signal-to-background ratios for measurement of J/ψ and Υ with CMS detector show the excellent capability to study the heavy-quarkonia cross-sections versus centrality, rapidity y and transverse momentum p_T, in Pb+Pb collisions at $\sqrt{s_{NN}} = 5.5$ TeV, via their dimuon decay channel. The differential cross-sections analyses will help significantly to clarify the physics mechanisms of quarkonia states production in high-energy nucleus-nucleus collisions.

1. Introduction

The suppression of heavy quark-antiquark bound states with increasing energy density due to Debye screening of the colour potential in the plasma is generally agreed to be one of the most direct probes of Quark-Gluon-Plasma formation[1] .

The CERN SPS results[2,3] showed a strong anomalous suppression of J/ψ production in Pb+Pb collisions at $\sqrt{s_{NN}} = 17.3$ GeV. The further study of J/ψ production in detail at $\sqrt{s_{NN}} = 200$ GeV in Au+Au and Cu+Cu collisions at RHIC revealed the global J/ψ suppression ammounts in factor three between most central and peripheral collisions[4] which can be explained only partially by calculations including shadowing and absorption[5] . As for Υ production, the cross-section is large enough to be observed at RHIC, with limited statistics, its suppression is not expected until the high initial temperatures foreseen at LHC are reached.

The recent theoretical analysis[6] shows that the direct J/ψ could survive for temperature as high as 1.5 Tc (The critical temperature for the phase transition is about 200 MeV) which could be out of the range of the RHIC. At the LHC energies a possible regeneration of the J/ψ, due to the high production of c-bar c pairs, is expected and may compensate an anomalous suppression[7] .

Having large muon detector acceptance and high precisions tracker, the CMS detector is particularly well suited to study the quarkonia state production in the dimuon channel.

Over than 2 orders of magnitute for J/ψ and 4 orders of magnitude for Υ separate the production cross-sections of resonances decaying into two muons from the

Pb+Pb inelastic cross-section. A complete and statistically significant simulation of the quarkonia detection using an event generator like HIJING[8] to simulate a Pb+Pb collision and the official tools of the CMS Collaboration for the tracking of secondaries based on GEANT4[9] would need an unavailable calculation power. Therefore, a combination of fast and slow simulations is used in this analysis. The detailed (full) GEANT4 simulation was used to provide the single particles trigger and reconstruction efficiencies and mass resolution with a few 10^7 particles of different types relevant to the muon signal and background (muons, pions, kaons, $b-, c-$ hadrons). The corresponding detector response functions (trigger acceptances, mass resolutions, reconstruction efficiencies, etc) are parametrized with detailed simulations for the each type of particle[10] . The obtained parameterizations being applied to heavy ion events are used in a fast Monte Carlo to produce the finally corrected yields of detected particles (Υ , J/ψ and background contribution) depending on dimuon mass, p_T and η[10] .

2. CMS detector

A detailed description of the detector elements can be found in the corresponding Technical Design Reports[11–14] .

The CMS detector is designed to identify and measure muons, electrons, photons and jets over a large energy and rapidity range. The CMS is particularly well suited to study the Υ and J/ψ families, the continuum up to the Z^0 mass and higher masses through dimuon decay channel. The impact parameter (centrality) of the collision can be determined from measurements of transverse energy production over the range $|\eta| < 5$.

The central element of CMS is the magnet, a 13 m long, 6 m diameter, high-field solenoid (a uniform 4 T field) with an internal radius of $\approx$3 m. The hadronic (HCAL) and electromagnetic (ECAL) calorimeters are located inside the coil (except the forward calorimeter) and cover (including the forward calorimeter) from -5 to 5 pseudorapidity units. The HF calorimeter covers the region $3 < |\eta| < 5$.

The overall number of the nuclear interaction length before penetrating muon stations (λ) is 11-16 over $|\eta| < 3$[15] . Muons loose 3 GeV in the calorimeter due-to ionization losses. The probability for hadron to not interact in calorimeter at al is 0.005%. The first absorber, the electromagnetic calorimeter, is 1.3 m from the interaction point, eliminating a large fraction of the hadronic background.

The tracker covers the pseudorapidity regions $|\eta| < 2.5$. Starting from the beam axis the tracker is composed of two different types of detectors: pixels and silicon strips. The pixel detector consists of 3 barrel layers located at 4,7,11 cm from the beam axis with granularity $150 \times 150\mu\text{m}^2$ and 2 forward layers with granularity $150 \times 300\mu\text{m}^2$ located at the distances of 34 and 43 cm in z from the center of detector.

Silicon strip detectors are divided into inner and outer sections and fill the tracker area from 20 cm to 110 cm (10 layers) in the transverse direction and up to

260 cm (12 layers) in longitudinal direction.

The CMS muon stations cover the pseudorapidity regions $|\eta| < 2.4$ and consist of drift tube chambers (DT) in the barrel region (MB), $|\eta| < 1.2$, cathode strip chambers (CSCs) in the endcap regions (ME), $0.9 < |\eta| < 2.4$, and resistive plate chambers (RPCs) in both barrel and endcaps, for $|eta| < 2.1$. The RPC detector is dedicated to triggering, while the DT and CSC detectors, used for precise momentum measurements, also have the capability to self-trigger up to $|\eta| < 2.1$.

3. Signal and background processes

Due-to current theoretical ambiguities the shadowing is the only effect taken into account in the events simulations. Dense matter effects like suppression or modification in the high Pt hadron spectra[16] are not considered.

3.1. *Quarkonia production rates*

The quarkonium cross sections per nucleon in Pb+Pb interactions are considered in the color evaporation model[17–19] with the MRST parton densities weighted by the EKS98 parameterisation [20] of nuclear shadowing effects. The parameters (PDF functions and scales) and the model prescription are indicated in[17] at page 258-259. The PbPb cross sections are obtained by scaling the per nucleon cross section with A^2, where $A = 208$ for Pb. The calculations include nuclear shadowing, but no additional effect like the absorption is taken into account, since the interplay of shadowing and nucleon absorption as a function of $\sqrt{s}$ is not known.

Table 1 shows the inclusive quarkonia production cross section values averaged over impact parameter and multiplied by the corresponding dimuon branching ratios in the case of Pb+Pb collisions.

Table 1. Inclusive cross sections for quarkonium production in minimum bias Pb+Pb collisions multiplying by decay branching ratio into $\mu^+\mu^-$.

$B_{\mu\mu}\ \sigma_{prod}$ (μb)				
J/ψ	ψ'	Υ	Υ'	Υ''
48930	879	304	78.8	44.4

The kinematical distributions in p_T and η of the quarkonia are generated according to[17] .

3.2. *Muon background rates*

The two main sources of background in the dimuon invariant mass spectrum are:

- Combinatorial muon pairs from the decays of **charged pions and kaons**.

The soft hadrons multiplicity at impact parameter b=0 is adjusted to the value expected at the LHC. Although extrapolations from RHIC suggest $dN^{\pm}/d\eta|_{\eta=0}$ as low as 2000[21] , two hypotheses with higher multiplicity values at η= 0 represented 5% most central events are used:
a) high multiplicity value with $dN^{\pm}/d\eta$ =5000
b) low multiplicity with $dN^{\pm}/d\eta$ =2500
In the fast Monte Carlo, only pions and kaons, which all together represent about 90% of all charged particles emitted in the collision, are considered. Neither neutral hadrons nor gammas are taken into account.
Protons can be registered as muons in muon chambers either due-to punchthrough of hadronic shower in the CMS calorimeter (11-16 λ in barrel and 11 λ in endcap)[15,22] or via the direct penetration of surviving protons into muon stations. The punchthrough in the muon chambers due-to hadronic shower in the CMS calorimeter is less then 1.2×10^{-5} for hadrons with $p_T < 10$ GeV/c[23] . The probability of the direct penetration of the surviving proton into muon stations is lower than 1.6×10^{-5}.
For pions and kaons, the overall probability (punchthrough and decay into muon) to be registered is about 0.003 and 0.01 correspondingly[10] and is more than two order of magnitude higher then the probability to register proton. Therefore, protons are neglected in the present study.
A ratio K/π=11% and kinematical spectra are extracted from HIJING. Pions and kaons have quite different $p_{\rm T}$ distributions: the mean $p_{\rm T}$ value for pions is 0.44 GeV/c while for kaons it is equal to 0.6 GeV/c . This fact explains why in spite of the K/π ratio of 11% at the generation level this ratio grows up to 80% at the detector level.

- The other important source of background are the muons coming from the **open $Q\overline{Q}$ pair** production produced in individual NN collisions. The number of pairs produced in Pb+Pb collisions as a function of impact parameter b is

$$N(Q\overline{Q}) = \sigma(Q\overline{Q})T_{AA}(b) ,$$

directly proportional to the nuclear overlap $T_{AA}(b)$ where $T_{\rm PbPb}(0)$ = 30.4/mb. The NLO $Q\overline{Q}$ production cross sections in 5.5 TeV pp interactions are $\sigma(c\bar{c})$= 7.5 mb and $\sigma(b\bar{b})$= 0.2 mb. These cross sections do not include the shadowing effect, the reducing factor of which is taken to be 35% and 15% respectively[24] . The number of $Q\overline{Q}$ pairs at $b = 0$ for Pb+Pb collisions, without and with the shadowing effect, are assumed to be the same for both sets of π/K multiplicity[10] .
Each $Q\overline{Q}$ has a probability to give one or more muons in the final state. These probabilities for $c\bar{c}$ are 0.819, 0.171, 0.010 for zero, one and more then two muons correspondingly. For $b\bar{b}$ the corresponding vaues are 0.626, 0.309, 0.065. They were obtained from Pythia 6.205 event generator. The muon $p_{\rm T}$ and η distributions are extracted from[17] . The large difference in

the average $p_{\rm T}$ values $\langle p_{\rm T}^{\mu}\rangle_c = 0.55$ GeV/c and $\langle p_{\rm T}^{\mu}\rangle_b = 1.45$ GeV/c counterbalances the effect of the much larger number of $c\bar{c}$ pairs at the generation level.

The choice of the impact parameter, b, governs the overall charged particles multiplicity, N_{ch}, as well as the number of open heavy quark pairs, $N_{Q\overline{Q}}$ produced in the collision. Charged particles and open charm and bottom production give the essential contributions to the dimuon background.

3.3. *Reconstruction and analysis*

3.3.1. *Dimuon trigger efficiency and acceptance*

The response of the CMS detector to muons, punchthrough pions and kaons is parameterised by 2-dimensional (p, η) acceptance and trigger tables. The single particles are fully tracked and digitized in detector using CMS detailed simulation[9] . Each track is accepted or rejected according to the Level-1 and Level-2 heavy-ion dimuon trigger criteria (i.e. the standard L1 *pp* muon trigger, with a low-quality μ condition and without pt-cut, and the standard *pp* L2 muon trigger, see Ref.[22] . The corresponding efficiencies, $\epsilon_{\rm trig}^{L1}(p, \eta)$ and $\epsilon_{\rm trig}^{L2}(p, \eta)$, are then computed. The trigger efficiencies are of the order of 90% for the muons reaching the muon chambers. Pions and kaons are detected through their decay into muon or punchthrough with corresponding efficiencies 0.3 % and 1% as indicated in Sect. 3.2.

The J/ψ and Υ acceptances are shown as a function of $p_{\rm T}$ in Fig. 1 for two η ranges: full detector ($|\eta| < 2.4$) and central barrel ($|\eta| < 0.8$). Because of their relatively small mass, low momentum J/ψ's ($p \lesssim 4$GeV/c) are mostly not accepted: their decay muons do not have enough energy to traverse the calorimeters and coil, and are absorbed before reaching the muon chambers. The J/ψ acceptance increases with $p_{\rm T}$, flattening out at $\sim$ 15% for $p_{\rm T} \gtrsim 12$GeV/c. The Υ acceptance starts at $\sim$40% at $p_{\rm T} = 0$ GeV/c and remains constant at $\sim$ 15% (full detector) or 5% (barrel only) for $p_{\rm T} > 4$ GeV/c. The $p_{\rm T}$-integrated acceptance is about 1.2% for the J/ψ and 26% for the Υ, assuming the input theoretical distributions (Sect. 3.1).

3.3.2. *Dimuon reconstruction efficiency and purity*

The dimuon reconstruction algorithm starts from L1 and L2 muon trigger candidates. The propagation is done from outside (Muon chambers) towards event primary vertex[10,25] . The final fit of the full trajectories is performed with a Kalman-fitter. The efficiency of a given muon pair is $\epsilon_{\rm pair}(p, \eta) = \epsilon_{\rm track1} \times \epsilon_{\rm track2} \times \epsilon_{\rm vertex}$. The dependence of the Υ reconstruction efficiency and purity on the event charged-particle multiplicity was obtained from a full GEANT4 simulation using the Υ signal dimuons embedded in HIJING Pb+Pb events. Note that particle multiplicities are given here in terms of *rapidity* densities. In the central barrel, the dimuon reconstruction efficiency remains above 80% for all multiplicities whereas the purity decreases slightly with increasing dN/dy but also stays above 80% even at multiplicities as

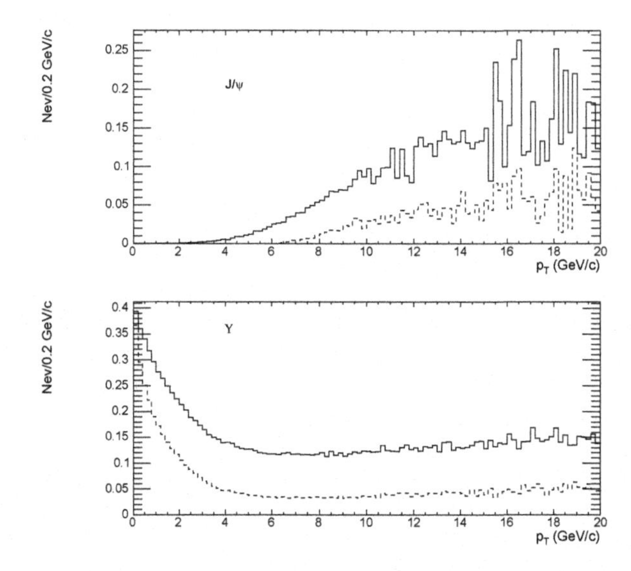

Fig. 1. J/ψ (top) and Υ (bottom) acceptances (convoluted with trigger efficiencies) as a function of p_T, in the full detector (barrel and endcap, $|\eta| < 2.4$, full line) and only in the barrel ($|\eta| < 0.8$, dashed line).

high as $dN/dy = 6500$. If (at least) one of the muons is detected in the endcaps, the efficiency and purity drop due to stronger reconstruction cuts. Nevertheless, for the $dN/d\eta \approx 2000$ multiplicity realistically expected in central Pb+Pb at LHC, the efficiency (purity) remains above 65% (90%) even including the endcaps.

3.3.3. *J/ψ and Υ mass resolutions*

At the Υ mass, the dimuon mass resolution for muon pairs in the central barrel, $|\eta| < 0.8$, is 54 MeV/c^2, as obtained from a Gaussian fit of the reconstructed $M_{\mu\mu}$ distribution (using a detailed MC simulation without background). In the *full* pseudorapidity range, the dimuon mass resolution is about 1% of the quarkonium mass: 35MeV/c^2 at the J/ψ mass and 86MeV/c^2 at the Υ mass. These dimuon mass resolutions (the best among the LHC experiments) provide a clean separation of the different quarkonia states. These values are used to smear the dimuon mass distribution in the fast MC studies. There is a slight dependence of the mass resolution on the event multiplicity. Increasing the multiplicity from $dN/dy = 0$ to 2500 degrades the mass resolution of the reconstructed Υ from 86 to 90 MeV/c^2. For larger particle densities, the resolution goes down more significantly because the endcap muons are then treated in a stricter way (stronger cuts are applied). This is done because the tracks in the endcap have up to two times worse momentum resolution than barrel tracks[22] . The efficiency of the forward muons is reduced but the purity is kept above 80%. The residual dependence of the mass resolution on the event multiplicity reflects the ratio between events with both muons in the barrel part of the tracker and events with at least one muon intersecting the endcap tracker disks. This ratio amounts to 0.25, 0.28 and 0.34 for $dN/dy = 0$, 2500 and 5000, respectively.

4. Results

In each Pb+Pb collision characterized by its impact parameter, all 5 resonances J/ψ ψ' and the 3 states of the Υ family, generated according to[17] , are superimposed to the background made of π's, K's and muons coming from open heavy quark pair production. The resonances are forced to decay into muons.
All muons which are within the acceptance of the CMS detector, i.e P > 2.8 GeV/c and $\mid \eta \mid< 2.4$ are recorded with the following quantities: momentum, pseudorapidity, charge and origin. The muons are weighted according to:

$$W_\mu = W_{origin} * W_{rec}$$

where W_{origin} is equal to 1 for the background muons and to $\sigma_{quarkonium}/\sigma_{PbPb}$ for the muons from quarkonia decay with σ(PbPb)= 8 barns. W_{rec} is the track reconstruction efficiency depending on the multiplicity, momentum, pseudorapidity and purity. Each muon candidate is also characterized by the trigger efficiencies W_{trig}^{LVL1} and W_{trig}^{LVL2}.
The invariant dimuon mass is calculated and smeared according to a Gaussian law with σ= 85 MeV/c^2 for masses above 5 GeV/c^2 and σ= 37 MeV for masses below 5 GeV/c^2 . One has to mention that introducing multiplicity dependant mass resolution do not affect the results[10] .

The weight of the pair is given by:

$$W(\mu_1\mu_2) = W_{\mu 1} * W_{\mu 2} * W_{trigger}$$

where $W_{trigger}$ is the combination of the trigger efficiencies of Level 1 and Level 2 of both muons. Since the muon charge is registered it is possible to separate like sign muon pairs from the opposite sign pairs.
This method allows to study $\approx 50\ 10^6$ Pb+Pb collisions at any impact parameter. Misidentification of muon charge by the low quality L1 muon trigger is not taking into account. For low momentum muon candidate with low quality assumption the misidentification of muon charge depends on η and may reach 10-15 % in the region $|\eta| > 2$. Level two muon trigger partially compensates the muon charge misidentification. After applying L3 dimuon trigger efficiency tables, the systematical shift between fast Monte-Carlo simulation and detailed simulation is less then 10 % and does not exceed systematical errors[10] .

The obtained dimuon mass distributions are then scaled to 0.5 nb^{-1}, corresponding to the Pb+Pb luminosity integrated in one month with average luminosity L $= 4 \times 10^{26}$ cm^{-2}s^{-1} and 50% machine operation efficiency. Figures 2 show the resulting opposite-sign dimuon mass distributions, for the high and low multiplicity cases and full acceptance ($|\eta| < 2.4$). The different quarkonia resonances appear on top of a continuum due to the various sources of decay muons: $\pi + K$, charm and bottom decays.

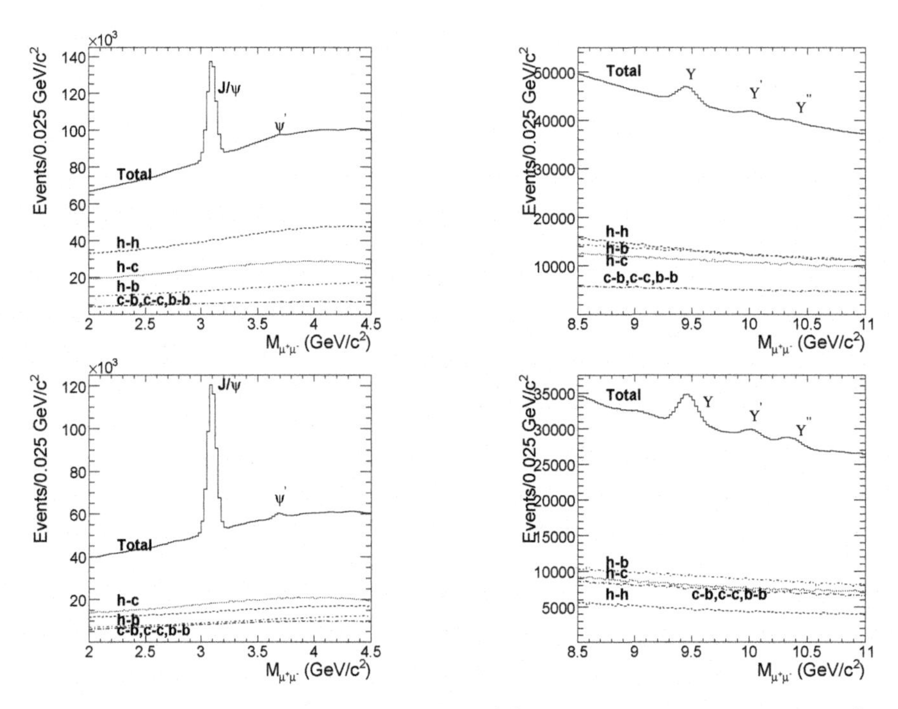

Fig. 2. Dimuon mass distributions measured within $|\eta| < 2.4$ for Pb+Pb events with $dN/d\eta = 5000$ (top) and $dN/d\eta = 2500$ (bottom) in the J/ψ (left) and Υ (right) mass regions. The main background contributions are also shown: h, c and b stand for $\pi + K$, charm, and bottom decay muons, respectively.

Assuming that the CMS trigger and acceptance conditions treat opposite-sign and like-sign muon pairs equally, the combinatorial like-sign background can be subtracted from the opposite-sign dimuon mass distribution, giving us a better access to the quarkonia decay signals[3] :

$$N^{\mathrm{Sig}} = N^{+-} - 2\sqrt{N^{++}\,N^{--}} \quad . \tag{1}$$

Figure 3 shows the *signal* dimuon mass distributions, after background subtraction, for two different scenarios: $dN/d\eta = 5000$ and $|\eta| < 2.4$ (worst case scenario); $dN/d\eta = 2500$ and $|\eta| < 0.8$ (best case scenario).

4.1. *Signal/Background ratio and statistics*

The statistics of J/ψ and Υ, $\Upsilon^{'}$ and $\Upsilon^{''}$ with both muons in $|\eta| < 2.4$ region expected in one month of data taking are 140000, 20000, 5900 and 3500 correspondingly for the multiplicity $dN/d\eta$=5000 and 180000, 25000, 7300, 4400 for the

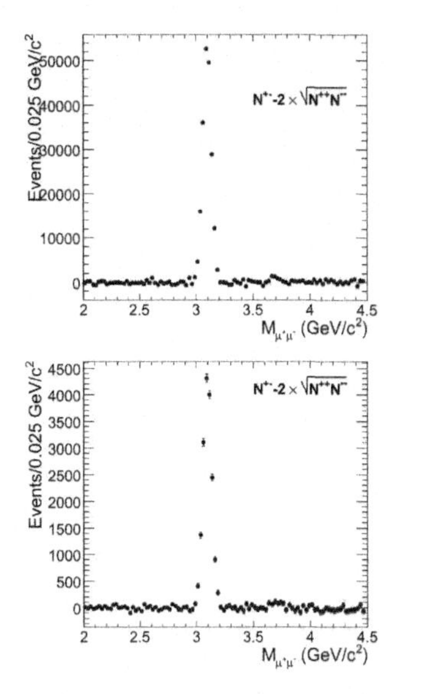

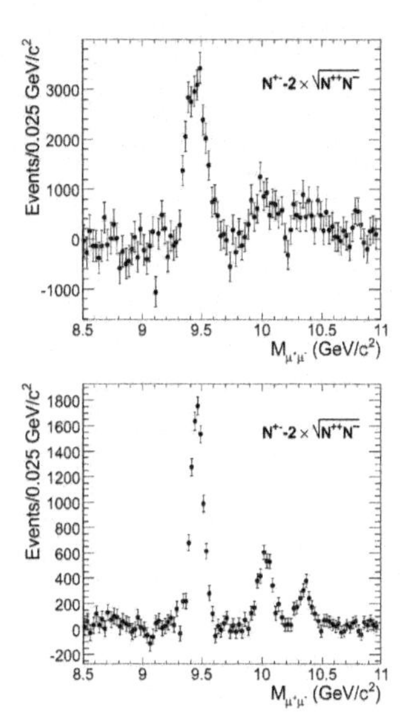

Fig. 3. Signal dimuon mass distributions in the J/ψ (left) and Υ (right) mass regions, as expected after one month of Pb+Pb running (0.5 nb^{-1}). Top panels for $dN/d\eta$ = 5000 and $|\eta| < 2.4$ (worst case scenario), bottom panels for $dN/d\eta$ = 2500 and $|\eta| < 0.8$ (best case scenario); assuming no quarkonia suppression.

multiplicity $dN/d\eta$=2500. The Signal/Background ratios are 0.6, 0.07 for J/ψ and Υ's for $dN/d\eta$=5000 correspondingly and 1.2, 0.12 for $dN/d\eta$=2500. The signal-to-background ratio (the number of events) collected in one month for the dimuons in J/ψ and Υ mass regions with both particles in $|\eta| < 0.8$ region are 2.75 (12600) and 0.52 (6000) for $dN/d\eta$=5000 and 4.5 (11600), 0.97 (6400) for $dN/d\eta$=2500. The background and reconstructed resonance numbers are in a mass interval $\pm\sigma$ where σ is the mass resolution.

These quantities have been calculated for an integrated luminosity of 0.5 nb^{-1} assuming an average luminosity $\mathcal{L} = 4 \times 10^{26}$ $cm^{-2}s^{-1}$ and a machine efficiency of 0.5. The expected statistics are large enough to allow further offline analysis for example in correlation with the centrality of the collision or the transverse momentum of the resonance.

One has to remark, that taking into account hot matter interactions and assuming the same magnitudes as in RHIC, the J/ψ could be suppressed by a factor of three (including cold matter effects, but excluding regeneration possibility), while the background (both from heavy and light quark production) should decrease by a factor of from two to five depending on p_T for the accepted kinematical range ($p >$ 3GeV/c, $|\eta| < 2.4$)[16] . Thus, the signal-to-backround ratio at the J/ψ mass should increase anyway.

4.1.1. *Transverse momentum and rapidity spectra*

The J/ψ transverse momentum and rapidity distributions are shown in Fig. 4, at the generated and reconstructed levels, for the two different multiplicity scenarios. The corresponding distributions for the Υ are shown in Fig. 5.

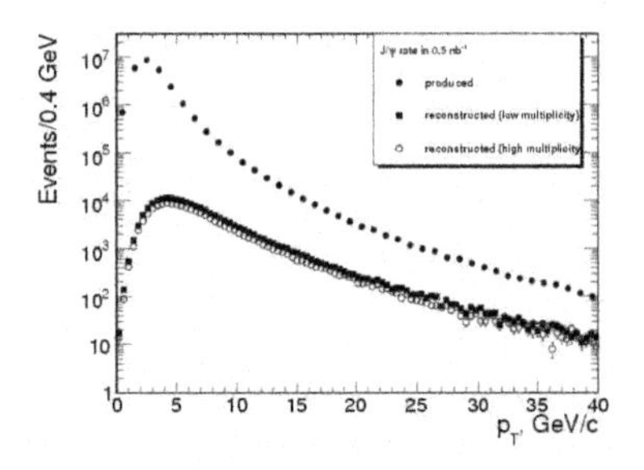

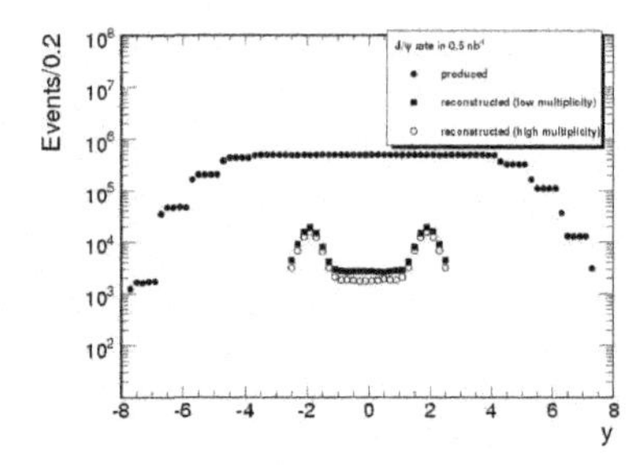

Fig. 4. p_T (left) and rapidity (right) distributions of the muon pairs in the J/ψ mass peak for PbPb at 5.5 TeV assuming no quarkonia suppression. The three distributions are the J/ψ's produced in 0.5 nb^{-1} (solid circles), and the reconstructed ones with either $dN/d\eta = 2500$ (squares) or $dN/d\eta = 5000$ (open circles).

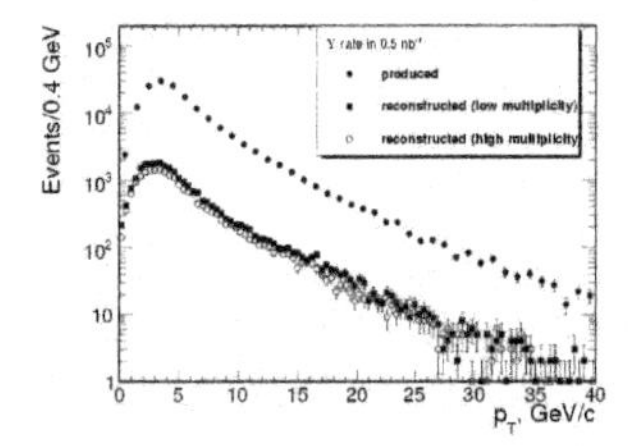

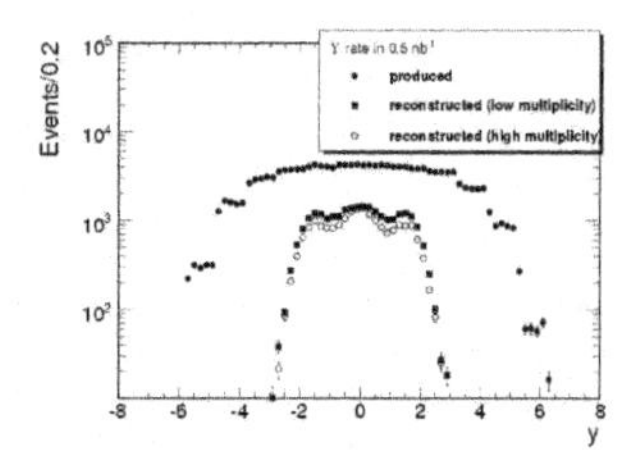

Fig. 5. p_T (left) and rapidity (right) distributions of the muon pairs in the Υ mass peak for PbPb at 5.5 TeV assuming no quarkonia suppression. The three sets of points correspond to: Υ's produced in 0.5 nb^{-1} (solid circles), reconstructed Υ's with $dN/d\eta$ =2500 (squares), and reconstructed Υ's with $dN/d\eta = 5000$ (open circles).

While the reconstructed Υ transverse momentum distributions have a shape quite similar to the generated one, a pronounced acceptance effect on the J/ψ spectrum is visible up to about 4 GeV/c, reflecting the J/ψ p_T acceptance curve (Fig. 1). The amount of statistics collected for the Υ resonance family in one nominal heavy-ion run with the high-level-trigger (HLT) settings discussed in[26] would allow one to study the p_T-dependence of the Υ' over Υ ratio, which is a very sensitive probe of the thermodynamical properties of the produced QGP[27] .

5. Discussion on the results and conclusions

With its $\approx 4\pi$ muon acceptance and full calorimetric coverage, CMS can make very significant and, in some respects, unique contribution to heavy ion physics. Studies of the Υ family, from pp to Pb+Pb , as well as from peripheral to central collisions, is likely to be of great interest at the LHC, just as the J/ψ has been for the SPS.

The key issue for CMS is the muon reconstruction efficiency in the tracker under conditions of extreme occupancies expected in Pb+Pb collisions. The reconstruction efficiency is in the range from 90% to 84% for dimuons with both muons in $|\eta| < 0.8$ for event multiplicity up to $dN_{\rm ch}/d\eta = 5000$. For muon pairs with at least one muon in the range $0.8 < |\eta| < 2.4$ the efficiency is 70% for multiplicity up to $dN_{\rm ch}/d\eta = 2500$ and falls down to 40 % for multiplicities above 2500 per unit of pseudorapidity. The purity is kept above 80 %. The Υ mass resolution is 56 MeV for the dimuons with both muons within pseudorapidity range $|\eta| < 0.8$. For the entire pseudorapidity region $|\eta| < 2.4$, the Υ mass resolution is 85 MeV while the J/ψ mass resolution is 35 MeV.

The large rapidity aperture of the muon detector, as well as the precise tracking, result in high statistics and a very good separation between the Υ states. The higher masses of the Υ states favour their measurement in the barrel with the muon momentum detection threshold limited to 3 GeV/c . Assuming only cold matter effect of shadowing, the number of events (20000 or 25000 for Υ depending on the multiplicity set and 140 000 or 170 000 for J/ψ) for one month of data taking (0.5 nb^{-1}) is high enough to provide a comparison between the several impact parameter bins and to carry out the study of the more differential analysis (dN/dy, dN/dp_T), which will contribute significantly to clarify the physics mechanisms behind the production (and destruction) of quarkonia states in high-energy nucleus-nucleus collisions.

References

1. T. Matsui and H. Satz, *Phys. Lett. B* **178**, 416 (1986).
2. M. C. Abreu *et al.* [NA50 Collaboration], *Phys. Lett. B* **450**, 456 (1999).
3. M. C. Abreu *et al.* [NA50 Collaboration], *Phys. Lett. B* **477**, 28 (2000).
4. H. Periera [PHENIX Collaboration] in *Proceedings of Quark Matter'05, Budapest, Hungary,August 2005.*
5. R. Vogt, nucl-th/0507027.
6. F. Karsch, D. Kharzeev and H. Satz, BNL-NT-05/50.
7. L. Grandchamp and R. Rapp, *Nucl. Phys. A* **715**, 545 (2003); hep-ph/0209141.
8. M. Gyulassy and X. N. Wang, *Phys. Rev. D* **44**, 3501 (1991).
9. OSCAR user's Guide at www.http://cmsdoc.cern.ch/swsi.html
10. O. Kodolova, M. Bedjidian, *J. Phys. G: Nucl. Part. Phys.* **34**, 143 (2007); http://stacks.iop.org/0954-3899/34/N143; CMS NOTE 2006/089.
11. HCAL Technical Design Report, CERN/LHCC 97-31, (1997).
12. MUON Technical Design Report, CERN/LHCC 97-32, (1997).
13. ECAL Technical Design Report,CERN/LHCC 97-33, (1997).
14. Tracker Technical Design Report,CERN/LHCC 98-6, (1998).
15. DAQ TDR CERN/LHCC/2002-26.
16. B. I. Abelev at al, nucl-ex/0606003 v3.
17. M. Bedjidian et al, in *Hard Probes in Heavy-Ion Collisions at the LHC, CERN-2004-*

009, hep-ph/0311048.
18. V. D. Barger, W. Y. Keung and R. J. Phillips, *Phys. Lett. B* **91**, 253 (1980).
19. V. D. Barger, W. Y. Keung and R. J. Phillips, *Z. Phys. C* **6**, 169 (1980).
20. K. J. Eskola, V. J. Kolhinen and C A. Salgado, *Eur. Phys. J. C* **9** (1999); hep-ph/9807297.
21. D. Kharzeev, E. Levin and M. Nardi, arXiv:hep-ph/0111315.
22. CMS Collaboration, in *The Physics Technical Design Report, part I, CERN-LHCC-2006-001.*
23. D. Chrisman and T. Moers, CMS TN/93-106.
24. *ALICE Physics Performance Report*, vol. II, CERN/LHCC 2005-030.
25. M. Bedjidian, O. Kodolova, S. Petrushanko, CMS NOTE 1999/004.
26. The CMS collaboration et al, CMS TDR Addendum: High-Density QCD with Heavy Ions, *J. Phys. G: Nucl. Part. Phys.* **34**, 2307 (2007); http://stacks.iop.org/0954-3899/34/2307.
27. J. F. Gunion and R. Vogt, R.,*Nucl. Phys. B* **492**, 301 (1997).

BOSE-EINSTEIN CORRELATIONS OF NEUTRAL GAUGE BOSONS IN *pp* COLLISIONS

G. A. KOZLOV

Bogolyubov Laboratory of Theoretical Physics
Joint Institute for Nuclear Research,
Joliot Curie st., 6, Dubna, Moscow region, 141980 Russia

The theory for Bose-Einstein correlations in case of neutral gauge bosons in *pp* collisions at high energies is presented. Based on quantum field theory at finite temperature the two-particle Bose-Einstein correlations of neutral gauge bosons are carried out for the first time. As a result, the important parameters of the correlation functions can be obtained for the Z^0Z^0 pairs. The correlations of two bosons in 4-momentum space presented in this paper offer useful and instructive complimentary viewpoints to theoretical and experimental works in multiparticle femtoscopy and interferometry measurements at hadron colliders.

1. Introduction

An investigation of the space-time extension or even squeezing of particle sources via the multiparticle quantum-statistics correlation in high energy interactions is still attract the attention of physical society in both experiment and theory. Over the past few decades, a considerable number of successful studies have been done in this direction[1] . It is well understood that the studies of correlations between produced particles, the effects of coherence and chaoticity, an estimation of particle emitting source size play an important role in high energy physics.

By studying the Bose-Einstein correlations (BEC) of identical particles (we mean like-sign charge particles and the neutral charge ones), it is possible experimentally to determine the time scale and spatial region over which particles do not have the interactions. Such a surface is called as decoupling one. In fact, for an evolving system such as *pp* collisions, it is not really a surface, since at each time there is a spread out surface due to fluctuations in the last interactions, and the shape of this surface evolve even in time. The particle source is not approximately constant because of energy-momentum conservation constraint.

More than half a century ago Hanbury-Brown and Twiss[2] used BEC between photons to measure the size of distant stars. In the works[3,4] , the master equations for evolution of thermodynamic system that can be created at the final state of a high multiplicity process were established. The equations have the form of the field operator evolution equation (Langevin-like[5]) and allows one to gain the basic features of the emitting source space-time structure. In particular, it has been conjectured and further confirmed that the size of the emitting source through BEC is strongly affected by non-classical off-shell effect.

The shapes of BEC function were experimentally established in the LEP experiments ALEPH[6] , DELPHI[7] and OPAL[8] , and ZEUS Collaboration at HERA[9] , which also indicated a dependence of the measured correlation radius on the hadron (π, K) mass. The results for $\pi^{\pm}\pi^{\pm}$ and $\pi^{\pm}\pi^{\mp}$ correlations with $p\bar{p}$ collisions at $\sqrt{s}$ = 1.8 TeV were published by E735 Collaboration in[10] .

The correlations between heavy particles (e.g., neutral gauge bosons Z^0Z^0) of Bose-Einstein type have not been carried out previously at hadron colliders. Such a study can be addressed to the Large Hadron Collider (LHC) which provide proton-proton interactions at $\sqrt{s} = 14$ TeV centre-of-mass system (c.m.s.) energy.

In this work, we make an attempt to demonstrate that the problem of properties of the genuine interactions can be explored using experimental data which can be collected by ATLAS and CMS Collaborations at the LHC. These data can be analyzed through the compared measures of some inclusive distributions and final state correlations.

One of the aims of this paper is to carry out the proposal for the experimental measurements of the Z^0Z^0 pair correlations.

This exploration will be theoretically supported by the quantum field theory at finite temperature (QFT_{β}) model approach[3] . It is known that the effective temperature of the vacuum or the ground state or even the thermalized state of particles distorted by external forces is occurring in models quantized in external fields. One of the main parameters of the model is the temperature of the particle source under the random source operator influence. The main channels are the di-lepton production $pp \to Z^0Z^0 \to 2e^-2e^+$, $2\mu^-2\mu^+$, $e^-e^+\mu^-\mu^+$ in pp collisions.

An efficient selection of leptons needs to be performed according to the following criteria. First, all leptons were required to lie in the pseudorapidity range covered by, e.g., the CMS muon system that is, $|\eta| \leq 2.4$. Second, the leptons were required to be unlikely charged in pairs. Note that the acceptances of another multipurpose detector ATLAS in the azimuthal angle and pseudorapidity are close to the respective parameters of CMS.

The dilepton channel is especially promising from the experimental point of view, since it is expected that the experimental facilities related for LHC (CMS and ATLAS detectors) will make it possible to record muons of energy in the TeV range with a resolution of about a few percent and an efficiency close to 100 %. Moreover, this channel is characterized by a maximum signal-to-background ratio in the energy region being considered.

2. BEC in case of two particles

A pair of identical bosons with the mass m produced incoherently (in ideal nondisturbed, noninteracting cases) from an extended source will have an enhanced probability $C_2(p_1,p_2) = N_{12}(p_1,p_2)/[N_1(p_1) \cdot N_2(p_2)]$ to be measured in terms of differential cross section σ, where

$$N_{12}(p_1, p_2) = \frac{1}{\sigma} \frac{d^2\sigma}{d\Omega_1\, d\Omega_2} \tag{1}$$

to be found close in 4-momentum space $\Re_4$ when detected simultaneously, as compared to if they are detected separately with

$$N_i(p_i) = \frac{1}{\sigma} \frac{d\sigma}{d\Omega_i}, \quad d\Omega_i = \frac{d^3\vec{p}_i}{(2\pi)^3\, 2E_{p_i}}, \quad E_{p_i} = \sqrt{\vec{p}_i^{\,2} + m^2}, \quad i = 1, 2. \tag{2}$$

On the other hand, the following relation can be used to retrieve the BEC function $C_2(Q)$:

$$C_2(Q) = \frac{N(Q)}{N^{ref}(Q)}, \tag{3}$$

where $N(Q)$ in general case refer to the numbers for neutral gauge bosons (eg., Z^0Z^0) with

$$Q = \sqrt{-(p_1 - p_2)_\mu \cdot (p_1 - p_2)^\mu} = \sqrt{M^2 - 4\,m^2}. \tag{4}$$

In definitions (3) and (4), N^{ref} is the number of particle pairs without BEC and $p_{\mu_i} = (\omega_i, \vec{p}_i)$ are four-momenta of produced bosons ($i = 1,\ 2$); $M = \sqrt{(p_1 + p_2)^2_\mu}$ is the invariant mass of the pair of bosons.

An essential problem in extracting the correlation is the estimate of the reference distribution $N^{ref}(Q)$ in Eq. (3). If there are other correlations beside the Bose-Einstein effect, the distribution $N^{ref}(Q)$ should be replaced by a reference distribution corresponding to the two-particle distribution in a geometry without BEC. Hence, the expression (3) represents the ratio between the number of Z^0Z^0 pairs $N(Q)$ in the real world and the reference sample $N^{ref}(Q)$ in the imaginary world. Note that the reference sample can not be directly observed in an experiment. Different methods are usually applied for the construction of reference samples[1] , however all of them have strong restrictions. One of the preferable methods is to construct the reference samples directly from data. For our aim for reference sample $N^{ref}(Q)$, it is suitable to use the pairs Z^0Z^0 from different (mixed) events.

It is commonly assumed that the maximum of two-particle BEC function $C_2(Q)$ is 2 for $\vec{p}_1 = \vec{p}_2$ if no any distortion and final state interactions are taking into account.

There are experimental difficulties in a determination of Z^0Z^0 correlations, which are associated with acceptance limitations and limited statistics in the Z^0Z^0 sample.

In general, the shape of the BEC function $C_2(Q)$ is model dependent. The most simple form of Goldhaber-like parameterization for $C_2(Q)$[11] has been used for data fitting:

$$C_2(Q) = C_0 \cdot (1 + \lambda e^{-Q^2R^2}) \cdot (1 + \varepsilon Q), \tag{5}$$

where C_0 is the normalization factor, λ is the so-called the chaoticity strength factor, meaning $\lambda = 1$ for fully incoherent and $\lambda = 0$ for fully coherent sources; the parameter R is interpreted as a radius of the particle source, often called as the "correlation radius", and assumed to be spherical in this parameterization. The linear term in (5) is supposed to be account within the long-range correlations outside the region of BEC. Note that distribution of bosons can be either far from isotropic, usually concentrated in some directions or almost isotropic, and what is important that in both cases the particles are under the random chaotic interactions caused by other fields in the thermal bath. In the parameterization (5) all of these problems are embedded in the random chaoticity parameter λ. To advocate the formula (5) it is assumed:
a. incoherent average over particle source where λ serve to account for:
- partial coherence,
- long-lived resonances associated with multiple distinguishable sources,
- $Z^0 Z^0$ purity;
b. spherical Gaussian density of particle emission cell (with radius R);
c. static source which means no time (energy) dependence.

In order to save the quantum pattern of particle production process and to avoid the static and undistorted character of particle emitter source we also suggest to use the $C_2(Q)$ function within QFT_β accompanying by quantum evolution approach in the form:

$$C_2(Q) = \xi(N) \cdot \left[1 + \frac{1}{(1+\alpha)(1+\alpha')}\,\tilde{\Omega}(Q) + \frac{2\sqrt{\alpha\alpha'}}{(1+\alpha)(1+\alpha')}\,\sqrt{\tilde{\Omega}(Q)}\right] \cdot F(Q, \Delta x), \tag{6}$$

where $\xi(N)$ depends on the multiplicity N as

$$\xi(N) = \frac{\langle N(N-1)\rangle}{\langle N\rangle^2}. \tag{7}$$

The function $F(Q, \Delta x)$ that expresses the correlation magnitude as a function of Q and two-particle relative distance Δx is a consequence of the Bogolyubov's principle of correlations weakening at large distances[12]

$$F(Q, \Delta x) = \frac{f(Q, \Delta x)}{f(p_1) \cdot f(p_2)} = 1 + r_f\, Q + \ldots \tag{8}$$

The function (8) is normalized as $F(Q, \Delta x = \infty) = 1$, and r_f is the measure of correlations weakening where $r_f \to 0$ as $\Delta x \to \infty$; $f(Q, \Delta x)$ is the two-particle distribution function with Δx, while $f(p_i)$ are one-particle probability functions with $i = 1, 2$.

The important parameter α (as well as α') in (6) summarizes our knowledge of other than space-time characteristics of the particle emitting source, and plays the role of a coherence parameter (see[4] for details).

The $\tilde{\Omega}(q)$ in (6) has the following structure in momentum space

$$\tilde{\Omega}(Q) = \Omega(Q) \cdot \gamma(n), \tag{9}$$

where

$$\Omega(Q) = \exp(-\Delta_{p\Re}) = \exp\left[-(p_1 - p_2)^{\mu}\, \Re_{\mu\nu}\, (p_1 - p_2)^{\nu}\right] \tag{10}$$

is the smearing smooth dimensionless generalized function, $\Re_{\mu\nu}$ is the (nonlocal) structure tensor of the space-time size (BEC formation domain), and it defines the spherically-like domain of emitted (produced) bosons.

To clarify with $\gamma(n)$ in (9) let us emphasize that most of experiments dealing with elementary particles at high energies are of an inclusive as one measures quantum effect of BEC on limited samples of particles produced only. The unobserved part of the rest particle system acts then as a kind of thermal (heat) bath influencing measured samples of data (observables). Actually, the temperature T being the most important parameter describing the influence of such a thermal bath is occurred in this model.

The function $\gamma(n)$ reflects the quantum thermal features of BEC pattern and is defined as

$$\gamma(n) = \frac{n^2(\bar{\omega})}{n(\omega)\, n(\omega')}, \quad n(\omega) \equiv n(\omega, \beta) = \frac{1}{e^{(\omega-\mu)\beta} - 1}, \quad \bar{\omega} = \frac{\omega + \omega'}{2}, \tag{11}$$

where $n(\omega, \beta)$ is the mean value of quantum numbers for Bose-Einstein statistics particles with the energy ω and the chemical potential μ in the thermal bath with statistical equilibrium at the temperature $T = 1/\beta$. The following condition $\sum_f n_f(\omega, \beta) = N$ is evident, where the discrete index f reflects the one-particle state f.

In terms of time-like R_0, longitudinal R_L and transverse R_T components of the space-time size R_μ the distribution $\Delta_{p\Re}$ looks like:

$$\Delta_{p\Re} \to \Delta_{pR} = (\Delta p^0)^2 R_0^2 + (\Delta p^L)^2 R_L^2 + (\Delta p^T)^2 R_T^2. \tag{12}$$

Seeking for simplicity one has ($R_L = R_T = R$)

$$\Delta_{pR} = (p_1^0 - p_2^0)^2 R_0^2 + (\vec{p}_1 - \vec{p}_2)^2 \vec{R}^2 \tag{13}$$

for identical bosons.

Hence, we have introduced a new parameter R_μ, a 4-vector, which defines the region of nonvanishing particle density with the space-time extension of the particle emission source. Expression (10) must be understood in the sense that $\Omega(Q)$ is a function that in the limit $R \to \infty$, strictly becomes a δ-function. For practical using with ignoring the energy-momentum dependence of α, and assuming that $\alpha' = \alpha$ (α is related with $C_2(0)$ and N), we get the expression with $\Omega(Q) \simeq \exp(-Q^2 R^2)$:

$$C_2(Q) \simeq \xi(N) \left\{1 + \lambda_{new}(\beta)\, e^{-Q^2 R^2} \left[1 + \lambda_{corr}(\beta)\, e^{+Q^2 R^2/2}\right]\right\}, \tag{14}$$

where the new intercept function becomes as $\lambda_{new} = \gamma(\omega, \beta)/(1 + \alpha)^2$, and the new coherence correction in the brackets of Eq. (14) carries an additional intercept function $\lambda_{corr} = 2\,\alpha/\sqrt{\gamma(\omega, \beta)}$. In fact, since $\alpha \neq \alpha'$ (because $\omega \neq \omega'$ and, therefore, the number of states identified here with the number of particles $n(\omega)$ with given energy is also different), one can use the general precise form (6) for C_2 with details given by Eqs. (9) and (11) and with α coherence function depending on the particle mass, the energy of final leptons produced in pairs within the decays of Z^0's and such characteristics of the emission process as the temperature T and chemical potential μ occurring in the definition of $n(\omega)$ in (11).

Since we did not follow special assumptions on the quantum operator level for $C_2(Q)$ from the initial stage, it may correspond to a physically real and observable effect at the LHC. This pattern may lead to a new squeezing state of correlation region.

3. Stochastic field and Green's function

Let us consider the stochastic field $B_\mu(x) = B_{\mu_{\tilde{s}}}(x,\tau)$ that depends on the arbitrary random source $\tilde{s}(x)$, and the fifth component τ means the "stochastic time". The differential equation of an evolution of the field operator $B_{\mu_{\tilde{s}}}(x,\tau) = B_\mu(x,\tau)$ in the system under the associated stochastic process is

$$\partial_\tau B_\mu(x,\tau) = O[B_\mu(x,\tau)]\,, \tag{15}$$

where $O[B_\mu(x,\tau)]$ is the differential stochastic operator which has the form

$$O[B_\mu(x,\tau)] = -\frac{1}{V}\frac{\delta J[B_\mu(x,\tau)]}{\delta B_\mu(x,\tau)} + \tilde{s}_\mu(x,\tau) \tag{16}$$

with a volume V being introduced by dimensional reason. The r.h.s. of Eq. (16) is the so-called stochastic operator where $J = \int d^4y L[B_\mu(y), \partial_\nu B_{\mu\nu}(y)]$ is the action defined by the Lagrangian density L; $\tilde{s}_\mu(x,\tau) = s_\mu(x,\tau) + n_\mu P$ carries the random stochastic history where the memory dissipation forces and the heat bath effects are included into $s_\mu(x) = s_\mu(x,\tau)$, the constant P emerges within the action of the stationary forces. Equation (15) is nothing other but the evolution equation of the Langevin type applied already to stochastic processes on the quantum operator level in derivation of multiparticle Bose-Einstein correlations[3] .

For simplicity, we assume that $\tilde{s}_\mu(x)$ varies stochastically with the Gaussian correlation function

$$\langle \tilde{s}_\mu(x)\tilde{s}_\nu(y)\rangle = const\,\delta_{\mu\nu}\exp(-z^2/l_{ch}^2)\,, \tag{17}$$

where $z_\nu = (x-y)_\nu$, and *const* is the strength of the noise described by the distribution function $\exp(-z^2/l_{ch}^2)$ with l_{ch} being the noise characteristic scale. Both *const* and l_{ch} define the influence of the (Gaussian) noise on, e.g., correlations between particles that "feel" an action of an environment. Actually, Eq. (15) can be transferred to the standard field equation of motion (in Euclidean space)

$$\frac{1}{V}\frac{\delta J[B_\mu(x)]}{\delta B_\mu(x)} = \tilde{s}_\mu(x) \tag{18}$$

with the source $\tilde{s}_\mu(x)$ if both B_μ and $\tilde{s}_\mu$ do not depend on "stochastic time" τ. In classical theory, the random process given by $\tilde{s}_\mu(x)$ is nothing other but the white (Gaussian) noise.

In this paper, we focus on the role of particle masses and energies, effects of coherence and distortion, and the heat bath influences which are rather important to describe the correlations between particles. To solve this problem, especially to derive the memory term in evolution equation one can use the general properties of QFT_β. The model is defined by the following generating functional in four-dimensional space-time

$$Z = \int DB_\mu \exp\left[-i \int d^4x L(B_\mu, B_{\mu\nu})\right] , \tag{19}$$

where

$$L = -\frac{1}{4} B_{\mu\nu} B^{\mu\nu} + \frac{1}{2}(m^2 + U) B_\mu B^\mu \tag{20}$$

with $B_{\mu\nu} = \partial_\mu B_\nu - \partial_\nu B_\mu$.

The direct calculations using the solution of Eq. (15) with the Lagrangian density (20) leads to the propagator of the field $B_\mu(x,\tau)$ distorted by $\tilde{s}_\mu(x,\tau)$. The transverse part of $B_\mu(x,\tau)$ will give the correct expression for the Euclidean vector field propagator at $\tau \to \infty$.

We are working with fields that correspond to a thermal field $B_\mu(x)$ with the standard definition of the Fourier transformed propagator $F[\tilde{G}_{\mu\nu}(p)]$

$$F[\tilde{G}_{\mu\nu}(p)] = G_{\mu\nu}(x-y) = Tr\left\{T[B_\mu(x) B_\nu(y)] \rho_\beta\right\} \tag{21}$$

with $\rho_\beta = e^{-\beta H}/Tr e^{-\beta H}$ being the density matrix of a local system in equilibrium at temperature T under the Hamiltonian H

$$H = \int \frac{d^3\vec{p}}{(2\pi)^3 2p^0} p^0 \sum_{\lambda=1}^{3} b^{\lambda^+}(p) b^\lambda(p) \tag{22}$$

with the operators of annihilation $b^\lambda(p)$ and creation $b^{\lambda^+}(p)$ to be defined later.

The interaction of $B_\mu(x)$ with the external field is given by the potential U. The equation of motion is

$$(\nabla^2 + m^2) B_\mu(x) = -J_\mu(x), \tag{23}$$

where $J_\mu(x) = U B_\mu(x)$ is the source density operator. A simple model like this allows one to investigate the origin of the unstable state of the thermalized equilibrium in a nonhomogeneous external field under the influence of source density operator $J_\mu(x) = U B_\mu(x)$. For example, the source can be considered as δ-like generalized function, $J_\mu(x) = \tilde{\mu}\, \rho(x,\epsilon) B_\mu(x)$ in which $\rho(x,\epsilon)$ is a δ-like succession giving the δ-function as $\epsilon \to 0$ (where $\tilde{\mu}$ is some massive parameter). This model is useful because the δ-like potential $U(x)$ provides the conditions for restricting the particle emission domain (or the deconfinement region). We suggest the following form:

$$J_\mu(x) = -J_{sys}(x)\, B_\mu(x) + J_{R_\mu}(x), \tag{24}$$

where the source $J_\mu(x)$ is a sum of a regular systematic motion part $J_{sys}(x)$ and the random source $J_{R_\mu}(x)$. The equation of motion (23) becomes

$$[\nabla^2 + m^2 - J_{sys}]B_\mu(x) = -J_{R_\mu}(x), \tag{25}$$

and the propagator satisfies the following equation (in the Fourier transformed form labeled by tilde):

$$[p_\mu^2 - m^2 + \tilde{J}_{sys}]\tilde{G}_{\mu\nu}(p_\mu) = \tilde{d}_{\mu\nu}(p), \tag{26}$$

where

$$d_{\mu\nu}(x) = \left(g_{\mu\nu} + \frac{1}{m^2}\frac{\partial^2}{\partial x_\mu \partial x_\nu} \right) \delta(x). \tag{27}$$

As the standard point, the Green's function of the vector field can be obtained from the one of the scalar field acting by the relevant operator $g_{\mu\nu} + m^{-2}\partial^2/(\partial x_\mu \partial x_\nu)$.

The solution of Eq. (23) is

$$B_\mu(x) = -\int dy\, G_{\mu\nu}(x,y)\, J_{R_\nu}(y), \tag{28}$$

where the Green's function obeys the Eq. (26).

4. Green's function and kernel operator

Let us go to the thermal field operator $B_\mu(x)$ by means of the linear combination of the frequency parts $B_\mu^1(x)$ and $B_\mu^{2^+}(x)$

$$B_\mu(x) = B_\mu^1(x) + B_\mu^{2^+}(x) \tag{29}$$

with $[B_\mu(x), B_\nu(y)] = iD_{\mu\nu}(x-y)$ and

$$B_\mu^1(x) = \int \frac{d^3\vec{p}}{(2\pi)^3 2(\vec{p}^2 + m^2)^{1/2}} \sum_{\lambda=1}^{3} \epsilon_\mu^{(\lambda)}(p)\tilde{b}^{(\lambda)}(p)\, e^{-ipx},$$

$$B_\mu^{2^+}(x) = \int \frac{d^3\vec{p}}{(2\pi)^3 2(\vec{p}^2 + m^2)^{1/2}} \sum_{\lambda=1}^{3} \epsilon_\mu^{(\lambda)}(p)\tilde{b}^{(\lambda)^+}(p)\, e^{ipx}.$$

The following properties of polarization vectors are the standard ones:

$$\epsilon_\mu^{(\lambda)}(p)\epsilon_\mu^{(\lambda')}(p) = g_{\lambda\lambda'},$$

$$\sum_{\lambda=1}^{3} \epsilon_\mu^{(\lambda)}(p)\epsilon_\nu^{(\lambda)}(p) = -g_{\mu\nu} + \frac{p_\mu p_\nu}{m^2}.$$

We assume that the deviation from the asymptotic free state given by the operator $a(\vec{p},t)$ is provided by the random operator $r(\vec{p},t): a(\vec{p},t) \to b(\vec{p},t) = a(\vec{p},t)+r(\vec{p},t)$. The operators $\tilde{b}^{(\lambda)}(p)$ and $\tilde{b}^{(\lambda)^+}(p)$ obey the following equations in $\Re_4$ (see details in[3]):

$$[\omega - \tilde{K}(p)]\tilde{b}^{(\lambda)}(p) = \tilde{F}(p) + \rho(\omega_P, \epsilon), \tag{30}$$

$$[\omega - \tilde{K}^+(p)]\tilde{b}^{(\lambda)^+}(p) = \tilde{F}^+(p) + \rho^*(\omega_P, \epsilon), \tag{31}$$

where $p_\mu = (\omega = p^0, \vec{p})$. Both equations (30) and (31) can be transformed into new equations for the frequency parts $B^1_\mu(x)$ and $B^{2^+}_\mu(x)$

$$i\partial_0 B^1_\mu(x) + \int_{\Re_4} K(x-y)\, B^1_\mu(y)dy = f_\mu(x), \tag{32}$$

$$-i\partial_0 B^{2^+}_\mu(x) + \int_{\Re_4} K^+(x-y)\, B^{2^+}_\mu(y)dy = f^+_\mu(x), \tag{33}$$

where

$$f_\mu(x) = \int \frac{d^3\vec{p}}{(2\pi)^3 2(\vec{p}^2+m^2)^{1/2}} \sum_{\lambda=1}^{3} \epsilon^{(\lambda)}_\mu(p)[\tilde{F}(p) + \rho(\omega_P,\epsilon)]e^{-ipx}, \tag{34}$$

$$f^+_\mu(x) = \int \frac{d^3\vec{p}}{(2\pi)^3 2(\vec{p}^2+m^2)^{1/2}} \sum_{\lambda=1}^{3} \epsilon^{(\lambda)}_\mu(p)[\tilde{F}^+(p) + \rho^*(\omega_P,\epsilon)]e^{ipx}. \tag{35}$$

The equations for field components $B^1_\mu(x)$ and $B^{2^+}_\mu(x)$ (32) and (33), respectively, are nonlocal within the presence of the formfactors $K(x-y)$ and $K^+(x-y)$, respectively. In principle, these formfactors can admit the description of locality for nonlocal interactions. At this stage, it must be stressed that we have new generalized evolution equations (32) and (33), which retain the general features of the propagating and interacting of the quantum vector fields with mass m that are in the heat bath (thermal reservoir) and are chaotically distorted by other fields. For further analysis, let us rewrite the system of Eqs. (32) and (33) in the following form:

$$i\partial_0 B^1_\mu(x) + K(x) \star B^1_\mu(x) = f_\mu(x), \tag{36}$$

$$-i\partial_0 B^{2^+}_\mu(x) + K^+(x) \star B^{2^+}_\mu(x) = f^+_\mu(x), \tag{37}$$

where $A(x) \star B(x)$ is the convoluted function of the generalized functions $A(x)$ and $B(x)$. Applying the direct Fourier transformation to both sides of Eqs. (36) and (37) with the following properties of the Fourier transformation

$$F[K(x) \star B^i_\mu(x)] = F[K(x)]\, F[B^i_\mu(x)] \;\; (i = 1, 2^+),$$

we get two equations

$$[p^0 + \tilde{K}(p)]\tilde{B}^1_\mu(p) = F[f_\mu(x)], \tag{38}$$

$$[-p^0 - \tilde{K}^+(p)]\tilde{B}_\mu^{2^+}(p) = F[f_\mu^+(x)]. \tag{39}$$

Finally, we have got the following equation for $\tilde{B}_\mu(p)$ field:

$$[-p^0 + \tilde{K}^+(p)][p^0 + \tilde{K}(p)]\tilde{B}_\mu(p) = \tilde{T}_\mu(p), \tag{40}$$

where

$$\tilde{T}_\mu(p) = [-p^0 + \tilde{K}^+(p)]F[f_\mu(x)] + [p^0 + \tilde{K}(p)]F[f_\mu^+(x)].$$

We are now at the stage of the main strategy: one has to identify the field $B_\mu(x)$ and the random source operator $J_{R_\mu}(x)$, introduced in Eq. (25, with the Fourier transformed field $\tilde{B}_\mu(p)$ and $\tilde{T}(p)$ in (40), respectively.

The next step is our requirement that Green's function $\tilde{G}_{\mu\nu}(p)$ in Eq. (26) and the function $\Gamma_{\mu\nu}(p)$, satisfying the equation

$$[-p^0 + \tilde{K}^+(p)][p^0 + \tilde{K}(p)]\tilde{\Gamma}_{\mu\nu}(p) = g_{\mu\nu} \tag{41}$$

must be equal to each other, i.e.

$$F[\tilde{G}_{\mu\nu}(p) - \tilde{\Gamma}_{\mu\nu}(p)] = 0.$$

The kernel operator $\tilde{K}(p)$ is

$$\tilde{K}(p) \simeq \epsilon\sqrt{1 + \frac{m^2}{\epsilon^2}}, \tag{42}$$

where $\epsilon = 2\sqrt{\vec{k}_l^2 + m_l^2}$ is the total energy of the final lepton-antilepton pair (with momentum $\vec{k}_l$ and the mass m_l for the lepton) produced within the decay of Z^0 boson being in the rest frame. To get $\tilde{K}(p)$ in the form(42) we used the fact that the full Green's function $\tilde{G}_{\mu\nu}(p)$ is given by the corresponding full Green's function of the scalar field[13] under the action by the differential operator $(g_{\mu\nu} - m^{-2}p_\mu p_\nu)$.

5. Source size

It has been emphasized[4] that there are two different scale parameters in the model considered here. One of them is the so-called "correlation radius" R introduced in (5) and (6) with (12). In fact, this R-parameter gives the pure size of the particle emission source without the external distortion and interaction coming from other fields. The other (scale) parameter is the stochastic scale L_{st} which carries the dependence of the particle mass, the α-coherence degree and what is very important — the temperature T-dependence:

$$L_{st} \simeq \left[\frac{1}{\alpha(N)\,|p^0 - \tilde{K}(p)|^2\, n(m,\beta)}\right]^{\frac{1}{2}} \to \left[\frac{1}{\alpha(N)\,4\vec{k}_l^2\,|1-\delta_k|^2\,\bar{n}(m,\beta)}\right]^{\frac{1}{5}}, \tag{43}$$

where

$$\delta_k = \sqrt{1 + \frac{m^2}{4\vec{k}_l^2}}$$

and the lepton mass m_l is neglected.

It turns out that the scale L_{st} defines the range of stochastic forces. This effect is given by $\alpha(N)$-coherence degree which can be estimated from the experiment within the two-particle BEC function $C_2(Q)$ when Q close to zero, $C_2(0)$, at fixed value of mean multiplicity $\langle N\rangle$:

$$\alpha(N) \simeq \frac{2-\bar{C}_2(0)+\sqrt{2-\bar{C}_2(0)}}{\bar{C}_2(0)-1}, \quad \bar{C}_2(0)=C_2(0)/\xi(N). \tag{44}$$

In formula (43), $\bar{n}(m,\beta)$ is the thermal relativistic particle number density

$$\bar{n}(m,\beta)=3\int\frac{d^3\vec{p}}{(2\,\pi)^3}\,n(\omega,\beta)=3\frac{\mu^2+m^2}{2\,\pi^2}\,T\sum_{l=1}^{\infty}\frac{1}{l}K_2\left(\frac{l}{T}\sqrt{\mu^2+m^2}\right), \tag{45}$$

where $K_2(...)$ is the modified Bessel function.

The coherence function α is another very important one that summarizes our knowledge of other than space-time characteristics of the particle emission source, and the prediction of α from an experiment is very instructive aim itself. For $\alpha=0$, one actually finds

$$1 < C_2(Q) < \xi(N)(1+\gamma e^{-Q^2R^2})$$

which is nothing other but the Goldhaber parameterization[11] with $0<\gamma<1$ being a free parameter adjusting the observed value of $C_2(Q=0)$.

Within our aim to explore the correlation between Z^0Z^0 the scale L_{st} has the form

$$L_{st}\simeq\left[\frac{e^{\sqrt{m^2+\mu^2}/T}}{12\,\alpha(N)\,\vec{k}_l^2\,(m^2+\mu^2)^{3/4}\left(\frac{T}{2\,\pi}\right)^{3/2}\left(1+\frac{15}{8}\frac{T}{\sqrt{m^2+\mu^2}}\right)|1-\delta_k|^2}\right]^{\frac{1}{5}}, \tag{46}$$

where the condition $l\,\beta\,\sqrt{m^2+\mu^2}>1$ for any integer l in (45) was taken into account. The only lower temperatures will drive L_{st} within formula (46) even if $\mu=0$ and $l=1$ with the condition $T<m$.

Note that the condition $\mu<m$ is a general restriction in the relativistic "Bose-like gas", and $\mu=m$ corresponds to the Bose-Einstein condensation.

For high enough T no μ - dependence is found for L_{st}:

$$L_{st}\simeq\left[\frac{\pi^2}{12\,\zeta(3)\,\alpha(N)\,\vec{k}_l^2\,T^3|1-\delta_k|^2}\right]^{\frac{1}{5}}, \tag{47}$$

where the condition $T>l\sqrt{m^2+\mu^2}, l=1,2,...$ is taken into account. The origin of formula (47) comes from

$$\bar{n}(m,\beta)\to\bar{n}(\beta)\simeq\frac{3\,T^3}{\pi^2}\,\zeta(3) \tag{48}$$

where neither a Z^0 boson mass nor the μ - dependence occurred; $\zeta(3)=\sum_{l=1}^{\infty}l^{-3}=1.202$ is the zeta-function with the argument 3.

To be close to the experiment there is necessary to include transverse momenta, where the Z^0 boson mass m, in Eqs. (45), (46), (47) is replaced by the transverse mass $m_T = \sqrt{m^2 + p_T^2}$.

Actually, the increasing of T leads to squeezing of the domain of stochastic force influence, and $L_{st}(T = T_0) = R$ at some effective temperature T_0. The higher temperatures, $T > T_0$, satisfy to more squeezing effect and at the critical temperature T_c the scale $L_{st}(T = T_c)$ takes its minimal value. Obviously $T_c \sim O(200\ GeV)$ defines the phase transition where the chiral symmetry restoration will occur. Since in this phase all the masses tend to zero and $\alpha \to 0$ at $T > T_c$ one should expect the sharp expansion of the region with $L_{st}(T > T_c) \to \infty$.

The qualitative relation between R and L_{st} above mentioned is the only one we can emphasize in order to explain the mass dependence of the source size.

6. Conclusions

To summarize: the theoretical proposal for two-particle Bose-Einstein correlation function in case of $Z^0 Z^0$ pairs in pp collisions is carried out for the first time.

The correlations of two bosons in 4-momentum space presented in this paper offer useful and instructive complimentary viewpoints to theoretical and experimental works in multiparticle femtoscopy and interferometry measurements at hadron colliders.

We find the time dependence of correlation function calculated in time-dependent external field provided by the operator $r(\vec{p}, t)$ and the chaotic coherence function $\alpha(m, \beta)$. The result can be compared with the static correlation functions (see, e.g.,[14] and the references therein mainly devoted to heavy-ion collisions) and also can be used for experimental data fitting.

The stochastic scale L_{st} decreases with increasing temperatures slowly at low temperatures, and it decreases rather abruptly when the critical temperature is approached.

Our results first predicted for correlation radius R are bothZ^0 boson mass and lepton energy dependent

$$R \sim \frac{e^{m/5T_0}}{\alpha^{1/5} |\vec{k}_l|^{2/5} m^{3/10} T_0^{3/10}} \tag{49}$$

for low values of $T_0 < m$, while for higher temperatures, $T_0 > \sqrt{m^2 + \mu^2}$, one has

$$R \sim \frac{1}{\alpha^{1/5} |\vec{k}_l|^{2/5} T_0^{3/5}}. \tag{50}$$

The theoretical correlation radius R at temperature T_0 decreases as Z^0-boson momentum increases. Both estimations (49) and (50) serve as the first approximation to explain the experimental data at different $\sqrt{s}$ and hence at T. We claim that the experimental measuring of R (in fm) can provide the precise estimation of the effective temperature T_0 which is the main thermal character in the $Z^0 Z^0$ pair emitter source (given by the effective dimension R) in the proper leptonic decaying channel $Z^0Z^0 \to l\bar{l}l\bar{l}$ with the final lepton energy $\sqrt{\vec{k}_l^2 + m_l^2}$ at given α fixed by $C_2(Q=0)$ and $\langle N \rangle$. Actually, T_0 is the true temperature in the region of multiparticle production with dimension $R = L_{st}$, because at this temperature it is exactly the creation of two particles (Z^0Z^0) occurred, and these particles obey the criterion of BEC.

References

1. R. M. Weiner, *Phys. Rep.* **327**, 249 (2000).
2. R. Hanbury-Brown and R.Q.Twiss, *Nature* **178** 1046 (1956).
3. G. A. Kozlov, O. V. Utyuzh and G. Wilk, *Phys. Rev.* **C68** 024901 (2003); G. A. Kozlov, *Phys. Rev. C* **58** 1188 (1998); *J. Math. Phys.* **42** 4749 (2001) and *New J. of Physics* **4** 23.1 (2002); G. A. Kozlov, "BEC and the particle mass", hep-ph/0512184; G. A. Kozlov, J. Elem. *Part. Phys. Atom. Nucl.* **36** 108 (2005); G. A. Kozlov, O. Utyuzh, G. Wilk and Z. Wlodarczyk, "Some forgotten features of the Bose-Einstein correlations", hep-ph/0710.3710.
4. G. A. Kozlov, "BEC and the particle mass", hep-ph/0512184.
5. P. Langevin, C. R. *Acad. Sci.* **146** 530 (1908).
6. The ALEPH Collab., *Eur. Phys. J. C* **36** 147 (2004); *Phys. Lett. B* **611** 66 (2005).
7. The DELPHI Collab., *Phys. Lett. B* **379** 330 (1996).
8. G. Abbiendi et al. (OPAL Collab.), *Phys. Lett. B* **559** 131 (2003).
9. The ZEUS Collab., *Phys. Lett. B* **583** 231 (2004).
10. T. Alexopoulos et al., [E735 Collaboration], *Phys. Rev. D* **48** 1931 (1993).
11. G. Goldhaber et al., *Phys. Rev. Lett.* **3** 181 (1959); G. Goldhaber et al., *Phys. Rev.* **120** 300 (1960).
12. N. N. Bogolyubov, "Quasiaveragies in problems of statistical mechanics", JINR report D-781, JINR, Dubna (1961).
13. G. A. Kozlov, "Bose-Einstein correlations and the stochastic scale of light hadrons emitter source", hep-ph/arXiv:0801.2072.
14. C. Y. Wong, W. N. Zhang, *Phys. Rev. C* **76** 034905 (2007).

BOSE-EINSTEIN CORRELATIONS AND THE STOCHASTIC SCALE OF LIGHT HADRONS EMITTER SOURCE

G. A. KOZLOV

Bogolyubov Laboratory of Theoretical Physics
Joint Institute for Nuclear Research,
Joliot Curie st., 6, Dubna, Moscow region, 141980 Russia

Based on quantum field theory at finite temperature we carried out new results for two-particle Bose-Einstein correlation (BEC) function $C_2(Q)$ in case of light hadrons. The important parameters of BEC function related to the size of the emitting source, mean multiplicity, stochastic forces range with the particle energy and mass dependence, and the temperature of the source are obtained for the first time. Not only the correlation between identical hadrons are explored but even the off-correlation between non-identical particles are proposed. The correlations of two bosons in 4-momentum space presented in this paper offer useful and instructive complimentary viewpoints to theoretical and experimental works in multiparticle femtoscopy and interferometry measurements at hadron colliders. This paper is the first one to the next opening series of works concerning the searching of BEC with experimental data where the parameters above mentioned will be retrieved.

1. Introduction

For the aim to explore the correlations of Bose-Einstein type (BEC) one needs to use the properties of a particle detector, e.g., its tracking system to study the hadron processes at some energy region. Such a study will be done soon in the next papers.

This paper describes an attempt to address the problems of BEC within the theoretical aspects prior the real data will be analyzed.

Over the past few decades, a considerable number of studies have been done on the phenomena of multi-particle correlations observed in high energy particle collisions (see the review in[1]). It is well understood that the studies of correlations between produced particles, the effects of coherence and chaoticity, an estimation of particle emitting source size and the temperature play an important role in this branch of high energy physics.

By studying the Bose-Einstein correlations of identical particles (e.g., like-sign charge particles of the same sort) or even off-correlations with respect to different-charge bosons, it is possible to predict and even experimentally determine the time and spatial region over which particles do not have the interactions. Such a surface is called as decoupling one. In fact, for an evolving system such as, e.g., $p\bar{p}$ collisions, it is not really a surface, since at each time there is a spread out surface due to fluctuations in the final interactions, and the shape of this surface evolve even in time. The particle source is not approximately constant because of energy-momentum conservation constraint.

More than half a century ago Hanbury-Brown and Twiss[2] used BEC between photons to measure the size of distant stars. In the papers in[3] and[4] , the master equations for evolution of thermodynamic system created at the final state of the (very) high multiplicity process were established. The equations have the form of the field operator evolution equation (Langevin-like[5]) that allows one to gain the basic features of the emitting source space-time structure. In particular, it has been conjectured and further confirmed that the BEC is strongly affected by non-classical off-shell effect.

The shapes of BEC function were experimentally established in the LEP experiments ALEPH[6] , DELPHI[7] and OPAL[8] , and ZEUS Collaboration at HERA[9] , which also indicated a dependence of the measured so-called correlation radius on the hadron (π, K) mass. The results for $\pi^{\pm}\pi^{\pm}$ and $\pi^{\pm}\pi^{\mp}$ correlations with $p\bar{p}$ collisions at $\sqrt{s} = 1.8$ TeV were published by E735 Collaboration in[10] .

One of the aims of this paper is to carry out the extended model of BEC in the framework of quantum field theory at finite temperature (QFT_{β}) approach which to be applied later to real experimental data on two-particle BEC. It is known that the effective temperature of the vacuum or the ground state or even the thermalized state of particles distorted by external forces is occurring in models quantized in external fields. One of the main parameters of the model considered here is the temperature of the particle source under the random source operator influence.

Among the results obtained in this paper we mention a theoretical estimate accessible to experimental measurements of two-particle BEC and proof that quantum-statistical evolution of particle-antiparticle correlations are not an artifact of the standard formalism but a quite general properties of particle physics. The effect (called as surprised one) for non-identical particles correlations was predicted already in[11] .

2. Two-particle BEC

A pair of bosons with the mass m produced incoherently (in ideal nondisturbed, noninteracting cases) from an extended source will have an enhanced probability $C_2(p_1, p_2) = N_{12}(p_1, p_2)/[N_1(p_1) \cdot N_2(p_2)]$ to be measured (in terms of differential cross section σ), where

$$N_{12}(p_1, p_2) = \frac{1}{\sigma} \frac{d^2\sigma}{d\Omega_1\, d\Omega_2} \tag{1}$$

to be found close in 4-momentum space $\Re_4$ when detected simultaneously, as compared to if they are detected separately with

$$N_i(p_i) = \frac{1}{\sigma} \frac{d\sigma}{d\Omega_i}, \quad d\Omega_i = \frac{d^3\vec{p}_i}{(2\pi)^3\, 2E_{p_i}}, \quad E_{p_i} = \sqrt{\vec{p}_i^{\,2} + m^2}, \quad i = 1, 2. \tag{2}$$

The following relation can be used to retrieve the BEC function $C_2(Q)$:

$$C_2^{ij}(Q) = \frac{N^{ij}(Q)}{N^{ref}(Q)}, \quad i, j = +, -, 0, \tag{3}$$

where $N^{ij}(Q)$ in general case refer to the numbers $N^{\pm\pm}(Q)$ for like-sign charge particles (eg., $\pi^\pm\pi^\pm$, $K^\pm K^\pm$, ...); $N^{\pm\mp}(Q)$ — for different charge bosons (eg., $\pi^\pm\pi^\mp$, $K^\pm K^\mp$, ...) or even for neutral charge particles $N^{00}(Q)$ (eg., $\pi^0\pi^0$, K^0K^0, ...) with

$$Q = \sqrt{-(p_1 - p_2)_\mu \cdot (p_1 - p_2)^\mu} = \sqrt{M^2 - 4\,m^2}. \tag{4}$$

In formula (3) and (4) N^{ref} is the number of pairs without BEC and $p_{\mu_i} (i = 1,\ 2)$ are four-momenta of produced particles, $M = \sqrt{(p_1 + p_2)^2_\mu}$ is the invariant mass of the pair of bosons. For reference sample, $N^{ref}(Q)$, the like-sign pairs from different events can be used. It is commonly assumed that the maximum of two-particle BEC function $C_2^{ii}(Q)$ is 2 for $\vec{p}_1 = \vec{p}_2$ if no any distortion and final state interactions are taking into account.

In general, the shape of BEC $C_2(Q)$ function is model dependent. The most simple form of Goldhaber-like parameterization for $C_2(Q)$[12] has been used for data fitting:

$$C_2(Q) = C_0 \cdot (1 + \lambda e^{-Q^2 R^2}) \cdot (1 + \varepsilon Q), \tag{5}$$

where C_0 is the normalization factor, λ is so-called the chaoticity strength factor, meaning $\lambda = 1$ for fully chaotic and $\lambda = 0$ for fully coherent sources; the parameter R is interpreted as a radius of the particle source, often called as the "correlation radius", and assumed to be spherical in this parameterization. The linear term in (5) is often supposed to be account for long-range correlations outside the region of BEC. However, the origin of these long-range correlations as well as the value of ϵ are unknown yet. Note that distribution of, e.g., pions and kaons can be far from isotropic, usually concentrated in narrow jets, and further complicated by the fact that the light particles with masses less than 1 GeV often come from decays of long-lived heavier resonances and also are under the random chaotic interactions caused by other fields in the thermal bath. In the parameterization (5) all of these problems are embedded in the random chaoticity parameter λ.

We obtained the $C_2(Q)$ function within QFT_β approach[3] in the form:

$$C_2(Q) = \xi(N) \cdot \left[1 + \frac{2\alpha}{(1+\alpha)^2} \sqrt{\tilde{\Omega}(Q)} + \frac{1}{(1+\alpha)^2} \tilde{\Omega}(Q)\right] \cdot F(Q, \Delta x), \tag{6}$$

where $\xi(N)$ depends on the multiplicity N as

$$\xi(N) = \frac{\langle N(N-1)\rangle}{\langle N\rangle^2}. \tag{7}$$

The consequence of the Bogolyubov's principle of weakening of correlations at large distances[13] is given by the function $F(Q, \Delta x)$ of weakening of correlations at large spread of relative position Δx

$$F(Q,\Delta x) = \frac{f(Q,\Delta x)}{f(p_1)\cdot f(p_2)} = 1 + r_f\, Q + \ldots \tag{8}$$

normalized as $F(Q,\Delta x=\infty)=1$. Here, $f(Q,\Delta x)$ is the two-particle distribution function with Δx, while $f(p_i)$ are one-particle probability functions with $i=1,2$; r_f is a measure of weakening of correlations with Δx: $r_f \to 0$ as $\Delta x \to \infty$.

The important parameter α in (6) summarizes our knowledge of other than space-time characteristics of the particle emitting source.

The $\tilde{\Omega}(Q)$ in (6) has the following structure in momentum space

$$\tilde{\Omega}(Q) = \Omega(Q)\cdot\gamma(n), \tag{9}$$

where

$$\Omega(Q) = \exp(-\Delta_{p\Re}) = \exp\left[-(p_1-p_2)^{\mu}\,\Re_{\mu\nu}\,(p_1-p_2)^{\nu}\right] \tag{10}$$

is the smearing smooth dimensionless generalized function, $\Re_{\mu\nu}$ is the (nonlocal) structure tensor of the space-time size (BEC formation domain), and it defines the spherically-like domain of emitted (produced) particles.

The function $\gamma(n)$ in (9) reflects the quantum features of BEC pattern and is defined as

$$\gamma(n) = \frac{n^2(\bar{\omega})}{n(\omega)\,n(\omega')},\quad n(\omega)\equiv n(\omega,\beta) = \frac{1}{e^{(\omega-\mu)\beta}-1},\quad \bar{\omega} = \frac{\omega+\omega'}{2}, \tag{11}$$

where $n(\omega,\beta)$ is the mean value of quantum numbers for BE statistics particles with the energy ω and the chemical potential μ in the thermal bath with statistical equilibrium at the temperature $T=1/\beta$. The following condition $\sum_f n_f(\omega,\beta)=N$ is evident, where the discrete index f reflects the one-particle state f.

Note that it is commonly assumed for a long time that there are no correlation effects among nonidentical particles (e.g., among different charged particles). This assumption is often used in normalizing the experimental data on C_2^{ii} with respect to C_2^{ij}. In the absence of interference or correlation effects between, e.g., π^+ and π^- mesons it is supposed that $C_2^{+-}=1$.

In terms of time-like R_0, longitudinal R_L and transverse R_T components of the space-time size R_μ the distribution $\Delta_{p\Re}^{ij}$ looks like $(i,j=+,-,0)$

$$\Delta_{p\Re}^{ij} \to \Delta_{pR}^{ij} = (\Delta p^0)^2R_0^2 + (\Delta p^L)^2R_L^2 + (\Delta p^T)^2R_T^2. \tag{12}$$

Seeking for simplicity one has $(R_L=R_T=R)$

$$\Delta_{pR}^{ii} = (p_1^0-p_2^0)^2R_0^2 + (\vec{p}_1-\vec{p}_2)^2\vec{R}^2 \tag{13}$$

for like-sign charge bosons, while

$$\Delta_{pR}^{ij} = (p_1^0+p_2^0)^2R_0^2 + (\vec{p}_1+\vec{p}_2)^2\vec{R}^2 \tag{14}$$

for different charge particles.

Obviously, the BEC effect with $\Omega^{ij} = \exp(-\Delta^{ij}_{pR})$ is smaller than that defined by $\Omega^{ii} = \exp(-\Delta^{ii}_{pR})$. The distribution Ω^{ij} gives rise to an off-correlation pattern between different charge particles. The evidence of C_2^{ij} correlation represents a quantum-statistical correlation between a particle and an antiparticle. Since we did not follow special assumptions on the quantum operator level for C_2 from the initial stage, it may correspond to a physically real and observable effect. This pattern may lead to a new squeezing state of correlation region. We obtain that within the QFT_β the BEC is more generally sensitive to particle-antiparticle correlations than it would be expected from the two-particle (symmetrized) wave function which never leads to such the correlations.

3. Green's function

In this paper, we would like to focus on the role of the particle mass, which influences the correlations between particles. To explore this problem, one must derive the memory history of evolution of particles produced in high energy collisions using the general properties of QFT at finite temperature.

We consider the thermal scalar complex fields $\Phi(x)$ that correspond to $\pi^\pm$ mesons with the standard definition of the Fourier transformed propagator $F[\tilde{G}(p)]$

$$F[\tilde{G}(p)] = G(x-y) = Tr\left\{T[\Phi(x)\Phi(y)]\rho_\beta\right\}, \tag{15}$$

with $\rho_\beta = e^{-\beta H}/Tre^{-\beta H}$ being the density matrix of a local system in equilibrium at temperature $T = \beta^{-1}$ under the Hamiltonian H.

We consider the interaction of $\Phi(x)$ with the external scalar field given by the potential U. In contrast to an electromagnetic field, this potential is a scalar one, but it is not a component of the four-vector. The Lagrangian density can be written

$$L(x) = \partial_\mu \Phi^\star(x)\partial^\mu \Phi(x) - (m^2 + U)\Phi^\star(x)\Phi(x)$$

and the equation of motion is

$$(\nabla^2 + m^2)\Phi(x) = -J(x), \tag{16}$$

where $J(x) = U\Phi(x)$ is the source density operator. A simple model like this allows one to investigate the origin of the unstable state of the thermalized equilibrium in a nonhomogeneous external field under the influence of source density operator $J(x)$. For example, the source can be considered as δ-like generalized function $J(x) = \tilde{\mu}\,\rho(x,\epsilon)\Phi(x)$ in which $\rho(x,\epsilon)$ is a δ-like succession giving the δ-function as $\epsilon \to 0$ (where $\tilde{\mu}$ is some massive parameter). This model is useful because the δ-like potential $U(x)$ provides the model conditions for restricting the particle emission domain (or the deconfinement region). We suggest the following form:

$$J(x) = -\Sigma(i\partial_\mu)\,\Phi(x) + J_R(x),$$

where the source $J(x)$ decomposes into a regular systematic motion part $\Sigma(i\partial_\mu)\,\Phi(x)$ and the random source $J_R(x)$. Thus, the equation of motion (16) becomes

$$[\nabla^2 + m^2 - \Sigma(i\partial_\mu)]\Phi(x) = -J_R(x),$$

and the propagator satisfies the following equation:

$$[-p_\mu^2 + m^2 - \tilde{\Sigma}(p_\mu)]\tilde{G}(p_\mu) = 1. \tag{17}$$

The random noise is introduced with a random operator $\eta(x) = -m^{-2}\,\Sigma(i\partial_\mu)$, for that the equation of motion looks like:

$$\{\nabla^2 + m^2[1 + \eta(x)]\}\Phi(x) = -J_R(x). \tag{18}$$

We assume that $\eta(x)$ varies stochastically with the certain correlation function (CF), e.g., the Gaussian CF

$$\langle \eta(x)\,\eta(y)\rangle = C\exp(-z^2\mu_{ch}^2), \;\; z = x - y,$$

where C is the strength of the noise described by the distribution function $\exp(-z^2/L_{ch}^2)$ with L_{ch} being the noise characteristic scale. Both C and μ_{ch} define the influence of the (Gaussian) noise on the correlations between particles that "feel" an action of an environment. The solution of Eq. (18) is

$$\Phi(x) = -\int dy\, G(x, y)\, J_R(y), \tag{19}$$

where the Green's function obeys the Eq.

$$\{\nabla^2 + m^2[1 + \eta(x)]\}G(x, y) = \delta(x - y).$$

The final aim might having been to find the solution of Eq. (19), and then average it over random operator $\eta(x)$. Note that the operator $M(x) = \nabla^2 + m^2[1 + \eta(x)]$ in the causal Green's function

$$G(x, y) = \frac{1}{M(x) + i\,o}\delta(x - y)$$

is not definitely positive. However, we shall formulate another approach, where the random force influence is introduced on the particle operator level.

We introduce the general non-Fock representation in the form of the operator generalized functions

$$b(x) = a(x) + r(x), \tag{20}$$

$$b^+(x) = a^+(x) + r^+(x), \tag{21}$$

where the operators $a(x)$ and $a^+(x)$ obey the canonical commutation relations (CCR):

$$[a(x), a(x')] = [a^+(x), a^+(x')] = 0,$$

$$[a(x), a^+(x')] = \delta(x - x').$$

The operator-generalized functions $r(x)$ and $r^+(x)$ in (20) and (21), respectively, include random features describing the action of the external forces.

Both b^+ and b obviously define the CCR representation. For each function f from the space $S(\Re_\infty)$ of smooth decreasing functions, one can establish new operators $b(f)$ and $b^+(f)$

$$b(f) = \int f(x)b(x)dx = a(f) + \int f(x)r(x)dx,$$

$$b^+(f) = \int \bar{f}(x)b^+(x)dx = a^+(f) + \int \bar{f}(x)r^+(x)dx.$$

The transition from the operators $a(x)$ and $a^+(x)$ to $b(x)$ and $b^+(x)$, obeying those commutation relations as $a(x)$ and $a^+(x)$, leads to linear canonical representations.

4. Evolution equation

Referring to[3] for details, let us recapitulate here the main points of our approach in the quantum case: the collision process produces a number of particles, out of which we select only one (we assume for simplicity that we are dealing only with identical bosons) and describe it by stochastic operators $b(\vec{p},t)$ and $b^+(\vec{p},t)$, carrying the features of annihilation and creation operators, respectively. The rest of the particles are then assumed to form a kind of heat bath, which remains in an equilibrium characterized by a temperature T (one of our parameters). We also allow for some external (relative to the above heat bath) influence on our system. The time evolution of such a system is then assumed to be given by a Langevin-type equation[3] for stochastic operator $b(\vec{p},t)$

$$i\partial_t b(\vec{p},t) = A(\vec{p},t) + F(\vec{p},t) + P \tag{22}$$

(and a similar conjugate equation for $b^+(\vec{p},t)$). We assume an asymptotic free undistorted operator $a(\vec{p},t)$, and that the deviation from the asymptotic free state is provided by the random operator $r(\vec{p},t)$: $a(\vec{p},t) \to b(\vec{p},t) = a(\vec{p},t) + r(\vec{p},t)$. This means, e.g., that the particle density number (a physical number) $\langle n(\vec{p},t)\rangle_{ph} = \langle n(\vec{p})\rangle + O(\epsilon)$, where $\langle n(\vec{p},t)\rangle_{ph}$ means the expectation value of a physical state, while $\langle n(\vec{p})\rangle$ denotes that of an asymptotic state. If we ignore the deviation from the asymptotic state in equilibrium, we obtain an ideal fluid. One otherwise has to consider the dissipation term; this is why we use the Langevin scheme to derive the evolution equation, but only on the quantum level. We derive the evolution equation in an integral form that reveals the effects of thermalization.

Equation (22) is supposed to model all aspects of the hadronization processes (or even deconfinement). The combination $A(\vec{p},t)+F(\vec{p},t)$ in the r.h.s of (22) represents the so-called *Langevin force* and is therefore responsible for the internal dynamics of particle emission, as the memory term A causes dissipation and is related to stochastic dissipative forces[3]

$$A(\vec{p},t) = \int_{-\infty}^{+\infty} d\tau K(\vec{p},t-\tau)b(\vec{p},\tau)$$

with $K(\vec{p},t)$ being the kernel operator describing the virtual transitions from one (particle) mode to another. At any dependence of the field operator K on the time, the function $A(\vec{p},t)$ is defined by the behavior of the system at the precedent moments. The operator $F(\vec{p},t)$ in (22) is responsible for the action of a heat bath of absolute temperature T on a particle in the heat bath, and under the appropriate circumstances is given by

$$F(\vec{p},t) = \int_{-\infty}^{+\infty} \frac{d\omega}{2\pi} \psi(p_\mu)\hat{c}(p_\mu)e^{-i\omega t}.$$

The heat bath is represented by an ensemble of coupled oscillators, each described by the operator $\hat{c}(p_\mu)$ such that $\left[\hat{c}(p_\mu), \hat{c}^+(p'_\mu)\right] = \delta^4(p_\mu - p'_\mu)$, and is characterized by the noise spectral function $\psi(p_\mu)^3$. Here, the only statistical assumption is that the heat bath is canonically distributed. The oscillators are coupled to a particle, which is in turn acted upon by an outside force. Finally, the constant term P in (22) (representing *an external source* term in the Langevin equation) denotes a possible influence of some external force. This force would result, e.g., in a strong ordering of phases leading therefore to a coherence effect.

The solution of equation (22) is given in $S(\Re_4)$ by

$$\tilde{b}(p_\mu) = \frac{1}{\omega - \tilde{K}(p_\mu)} [\tilde{F}(p_\mu) + \rho(\omega_P, \epsilon)], \tag{23}$$

where ω in $\rho(\omega,\epsilon)$ was replaced by new scale $\omega_P = \omega/P$. It should be stressed that the term containing $\rho(\omega_P,\epsilon)$ as $\epsilon \to 0$ yields the general solution to Eq. (22). Notice that the distribution $\rho(\omega_P,\epsilon)$ indicates the continuous character of the spectrum, while the arbitrary small quantity ϵ can be defined by the special physical conditions or the physical spectra. On the other hand, this $\rho(\omega_P,\epsilon)$ can be understood as temperature-dependent succession $\rho(\omega,\epsilon) = \int dx\, exp(i\omega - \epsilon)x \to \delta(\omega)$, in which $\epsilon \to \beta^{-1}$. Such a succession yields the restriction on the β-dependent second term in the solution (23), where at small enough T there is a narrow peak at $\omega = 0$.

From the scattering matrix point of view, the solution (23) has the following physical meaning: at a sufficiently outgoing past and future, the fields described by the operators $\tilde{a}(p_\mu)$ are free and the initial and the final states of the dynamic system are thus characterized by constant amplitudes. Both states, $\varphi(-\infty)$ and $\varphi(+\infty)$, are related to one another by an operator $S(\tilde{r})$ that transforms state $\varphi(-\infty)$ to state $\varphi(+\infty)$ while depending on the behaviour of $\tilde{r}(p_\mu)$:

$$\varphi(+\infty) = S(\tilde{r})\varphi(-\infty).$$

In accordance with this definition, it is natural to identify $S(\tilde{r})$ as the scattering matrix in the case of arbitrary sources that give rise to the intensity of $\tilde{r}$.

Based on QFT point of view, relation (20) indicates the appearance of the terms containing nonquantum fields that are characterized by the operators $\tilde{r}(p_\mu)$. Hence, there are terms with $\tilde{r}$ in the matrix elements, and these $\tilde{r}$ cannot be realized via real particles. The operator function $\tilde{r}(p_\mu)$ could be considered as the limit on an average value of some quantum operator (or even a set of operators) with an intensity that increases to infinity. The later statement can be visualized in the following mathematical representation:

$$\tilde{r}(p_\mu) = \sqrt{\alpha\,\Xi(p_\mu, p_\mu)},\ \Xi(p_\mu, p_\mu) = \langle \tilde{a}^+(p_\mu)\,\tilde{a}(p_\mu)\rangle_\beta,$$

where α is the coherence (chaotic) function that gives the strength of the average $\Xi(p_\mu, p_\mu)$.

In principal, interaction with the fields described by $\tilde{r}$ is provided by the virtual particles, the propagation process of which is given by the potentials defined by the $\tilde{r}$ operator function.

The condition $M_{ch} \rightarrow 0$ (or $\Omega_0(R) \sim \frac{1}{M_{ch}^4} \rightarrow \infty$) in the representation

$$\lim_{p_\mu \rightarrow p'_\mu} \Xi(p_\mu, p'_\mu) = \lim_{Q^2 \rightarrow 0} \Omega_0(R)\, n(\bar{\omega}, \beta) \exp(-q^2/2) \rightarrow \frac{1}{M_{ch}^4}\, n(\omega, \beta),$$

with

$$\Omega_0(R) = \frac{1}{\pi^2}\, R_0\, R_L\, R_T^2$$

means that the role of the arbitrary source characterized by the operator function $\tilde{r}(p_\mu)$ in $\tilde{b}(p_\mu) = \tilde{a}(p_\mu) + \tilde{r}(p_\mu)$ disappears.

5. Green's function and kernel operator

Let us go to the thermal field operator $\Phi(x)$ by means of the linear combination of the frequency parts $\phi^+(x)$ and $\phi^-(x)$

$$\Phi(x) = \frac{1}{\sqrt{2}}\,\left[\phi^+(x) + \phi^-(x)\right] \tag{24}$$

composed of the operators $\tilde{b}(p_\mu)$ and $\tilde{b}^+(p_\mu)$ as the solutions of equation (22) and conjugate to it, respectively:

$$\phi^-(x) = \int \frac{d^3\vec{p}}{(2\pi)^3 2(\vec{p}^2 + m^2)^{1/2}} \tilde{b}^+(p_\mu)\, e^{ipx},$$

$$\phi^+(x) = \int \frac{d^3\vec{p}}{(2\pi)^3 2(\vec{p}^2 + m^2)^{1/2}} \tilde{b}(p_\mu)\, e^{-ipx}.$$

The function $\Phi(x)$ obeys the commutation relation

$$[\Phi(x), \Phi(y)]_- = -iD(x)$$

with[14]

$$D(x) = \frac{1}{2\,\pi}\,\epsilon(x^0)\,\left[\delta(x^2) - \frac{m}{2\,\sqrt{x_\mu^2}}\,\Theta(x^2)\, J_1\left(m\sqrt{x_\mu^2}\right)\right],$$

where $\epsilon(x^0)$ and $\Theta(x^2)$ are the standard unit and the step functions, respectively, while $J_1(x)$ is the Bessel function. On the mass-shell, $D(x)$ becomes[14]

$$D(x) \simeq \frac{1}{2\pi}\,\epsilon(x^0)\left[\delta(x^2) - \frac{m^2}{4}\,\Theta(x^2)\right].$$

One can easily find two equations of motion for the Fourier transformed operators $\tilde{b}(p_\mu)$ and $\tilde{b}^+(p_\mu)$ in $S(\Re_4)$

$$[\omega - \tilde{K}(p_\mu)]\tilde{b}(p_\mu) = \tilde{F}(p_\mu) + \rho(\omega_P, \epsilon), \tag{25}$$

$$[\omega - \tilde{K}^+(p_\mu)]\tilde{b}^+(p_\mu) = \tilde{F}^+(p_\mu) + \rho^\star(\omega_P, \epsilon), \tag{26}$$

which are transformed into new equations for the frequency parts $\phi^+(x)$ and $\phi^-(x)$ of the field operator $\Phi(x)$ (24)

$$i\partial_0\phi^+(x) + \int_{\Re_4} K(x-y)\,\phi^+(y)dy = f(x) \tag{27}$$

$$-i\partial_0\phi^-(x) + \int_{\Re_4} K^+(x-y)\,\phi^-(y)dy = f^+(x), \tag{28}$$

where

$$f(x) = \int \frac{d^3\vec{p}}{(2\pi)^3\,(\vec{p}^2 + m^2)^{1/2}}[\tilde{F}(p) + \rho(\omega_P, \epsilon)]e^{-ipx},$$

and

$$f^+(x) = \int \frac{d^3\vec{p}}{(2\pi)^3\,(\vec{p}^2 + m^2)^{1/2}}[\tilde{F}^+(p) + \rho^\star(\omega_P, \epsilon)]e^{ipx}.$$

Here, the field components $\phi^+(x)$ and $\phi^-(x)$ are under the effect of the nonlocal formfactors $K(x-y)$ and $K^+(x-y)$, respectively. In general, these formfactors can admit the description of locality for nonlocal interactions.

At this stage, it must be stressed that we have new generalized evolution Eqs. (27) and (28), which retain the general features of the propagating and interacting of the quantum fields with mass m that are in the heat bath (reservoir) and are chaotically distorted by other fields. For further analysis, let us rewrite the Eqs. (27) and (28) in the following form:

$$i\partial_0\phi^+(x) + K(x) \star \phi^+(x) = f(x), \tag{29}$$

$$-i\partial_0\phi^-(x) + K^+(x) \star \phi^-(x) = f^+(x), \tag{30}$$

where $A(x) \star B(x)$ is the convoluted function of the generalized functions $A(x)$ and $B(x)$. Applying the direct Fourier transformation to both sides of Eqs. (29) and (30) with the following properties of the Fourier transformation

$$F[K(x) \star \phi^+(x)] = F[K(x)]F[\phi^+(x)],$$

we get two equations

$$[p^0 + \tilde{K}(p_\mu)]\tilde{\phi}^+(p_\mu) = F[f(x)], \tag{31}$$

$$[-p^0 + \tilde{K}^+(p_\mu)]\tilde{\phi}^-(p_\mu) = F[f^+(x)]. \tag{32}$$

Multiplying Eqs. (31) and (43) by $-p^0 + \tilde{K}^+(p_\mu)$ and $p^0 + \tilde{K}(p_\mu)$, respectively, we find

$$[-p^0 + \tilde{K}^+(p_\mu)][p^0 + \tilde{K}(p_\mu)]\tilde{\Phi}(p_\mu) = T(p_\mu), \tag{33}$$

where

$$T(p_\mu) = [-p^0 + \tilde{K}^+(p_\mu)]F[f(x)] + [p^0 + \tilde{K}(p_\mu)]F[f^+(x)].$$

We are now at the stage of the main strategy: we have to identify the field $\Phi(x)$ introduced in Eq. (15) and the field $\Phi(x)$ (24) built up of the fields ϕ^+ and ϕ^- as the solutions of generalized Eqs. (27) and (28). The next step is our requirement that Green's function $\tilde{G}(p_\mu)$ in Eq. (17) and the function $\Gamma(p_\mu)$, that satisfies Eq. (44)

$$[-p^0 + \tilde{K}^+(p_\mu)][p^0 + \tilde{K}(p_\mu)]\tilde{\Gamma}(p_\mu) = 1, \tag{34}$$

must be equal to each other, where the full Green's function $\tilde{G}(p^2, g^2, m^2)$

$$\tilde{G}(p_\mu) \to \tilde{G}(p^2, g^2, m^2) \simeq \frac{1 - g^2\,\xi(p^2, m^2)}{m^2 - p^2 - i\epsilon} \tag{35}$$

has the same pole structure at $p^2 = m^2$ as the free Green's function[14] with g being the scalar coupling constant and ξ is the one-loop correction of the scalar field. The dimensioneless function $1 - g^2\,\xi(p^2, m^2)$ is finite at $p^2 = m^2$.

We define the operator kernel $\tilde{K}(p_\mu)$ in (25) from the condition of the nonlocal coincidence of the Green's function $\tilde{G}(p_\mu)$ in Eq. (17), and the thermodynamic function $\tilde{\Gamma}(p_\mu)$ from (34) in $S(\Re_4)$

$$F[\tilde{G}(p_\mu) - \tilde{\Gamma}(p_\mu)] = 0.$$

We can easily derive the kernel operator $\tilde{K}(p_\mu)$ in the form

$$\tilde{K}^2(p) = \frac{m^2 + \vec{p}^2 - g^2\xi(p^2, m^2)\,{p^0}^2}{1 - g^2\xi(p^2, m^2)} \tag{36}$$

where[14]

$$\xi(m^2) = \frac{1}{96\,\pi^2\,m^2}\left(\frac{2\,\pi}{\sqrt{3}} - 1\right),\ p^2 \simeq m^2,$$

and

$$\xi(p^2, m^2) = \frac{1}{96\,\pi\,m^2}\left(i\sqrt{1 - \frac{4\,m^2}{p^2}} + \frac{\pi}{\sqrt{3}}\right),\ p^2 \simeq 4m^2.$$

The ultraviolet behaviour at $|p^2| >> m^2$ leads to

$$\xi(p^2, m^2) \simeq \frac{-1}{32\,\pi^2\,p^2}\left[\ln\frac{|p^2|}{m^2} - \frac{\pi}{\sqrt{3}} - i\,\pi\Theta(p^2)\right].$$

6. Stochastic forces scale

In paper[4] it has been emphasized that two different scale parameters are in the model which we consider here. One of them is the so-called "correlation radius" R introduced in (5) and (6) with (9) and (12), (13), (14). In fact, this R-parameter gives the pure size of the particle emission source without the external distortion and interaction coming from other fields. The other (scale) parameter is the stochastic scale L_{st} which carries the dependence of the particle mass, the α-coherence degree and what is very important — the temperature T-dependence:

$$L_{st} = \left[\frac{1}{\alpha(N) \left|p^0 - \tilde{K}(p)\right|^2 \bar{n}(m,\beta)} \right]^{\frac{1}{5}} . \tag{37}$$

It turns out that this scale L_{st} defines the range of stochastic forces acting the particles in the emission source. This effect is given by $\alpha(N)$-coherence degree which can be estimated from the experiment within the two-particle BE correlation function $C_2(Q)$ as Q close to zero, $C_2(0)$, at fixed value of mean multiplicity $\langle N \rangle$:

$$\alpha(N) \simeq \frac{2 - \bar{C}_2(0) + \sqrt{2 - \bar{C}_2(0)}}{\bar{C}_2(0) - 1}, \quad \bar{C}_2(0) = C_2(0)/\xi(N). \tag{38}$$

In formula (37) $\bar{n}(m,\beta)$ is the thermal relativistic particle number density

$$\bar{n}(m,\beta) = 3 \int \frac{d^3\vec{p}}{(2\,\pi)^3}\, n(\omega,\beta) = 3 \frac{\mu^2 + m^2}{2\,\pi^2}\, T \sum_{l=1}^{\infty} \frac{1}{l} K_2 \left(\frac{l}{T} \sqrt{\mu^2 + m^2} \right), \tag{39}$$

where K_2 is the modified Bessel function. For definite calculations we consider correlations between charge pions. The result can be extended to heavy particles case, e.g., for charge and neutral gauge bosons that is essential program for the LHC. The stochastic scale L_{st} tends to infinity in case of particles are on mass-shell, i.e., $|p^0 - \tilde{K}(p)| \to 0$ which enters the $L'_{st}s$ denominator (37). However, L_{st} will be bounded due to stochastic forces acting the particles where

$$\left|p^0 - \tilde{K}(p)\right|^2 \simeq \Delta\epsilon_p^2 = \epsilon_p^2 \left| \frac{p^0}{\epsilon_p} - 1 - \frac{g^2 \xi(p^2,m^2)}{2} \left(1 - \frac{{p^0}^2}{\epsilon_p^2} \right) \right|^2, \; \epsilon_p = \sqrt{m^2 + \vec{p}^2}$$

as $g^2 \xi(p^2, m^2) < 1$.

Within our aim to explore the correlation between charged pions, L_{st} has the form

$$L_{st} \simeq \left[\frac{e^{\sqrt{\mu^2+m^2}/T}}{3\,\alpha(N)\,\Delta\epsilon_p^2\,(\mu^2 + m^2)^{3/4} \left(\frac{T}{2\,\pi}\right)^{3/2} \left(1 + \frac{15}{8} \frac{T}{\sqrt{\mu^2+m^2}}\right)} \right]^{\frac{1}{5}}, \tag{40}$$

where the condition $l\,\beta\sqrt{m^2 + \mu^2} > 1$ for any integer l in (39) was taken into account. The only lower temperatures will drive L_{st} within the formula (40) even if $\mu = 0$ and $l = 1$ with the condition $T < m$. Note that the condition $\mu < m$ is a general restriction in the relativistic "Bose-like gas", and $\mu = m$ corresponds to the Bose-Einstein condensation.

For large enough T no the dependence of the chemical potential μ is found for L_{st}:

$$L_{st} \simeq \left[\frac{\pi^2}{3\,\zeta(3)\,\alpha(N)\,\Delta\epsilon_p^2\,T^3}\right]^{\frac{1}{5}}, \tag{41}$$

where the condition $T > l\sqrt{\mu^2 + m^2}, l = 1, 2, ...$ is taken into account. The origin of formula (41) comes from

$$\bar{n}(m,\beta) \to \bar{n}(\beta) = \frac{3\,T^3}{\pi^2}\,\zeta(3) \tag{42}$$

where neither a pion mass m- nor μ- dependence occurred; $\zeta(3) = \sum_{l=1}^{\infty} l^{-3} = 1.202$ is the zeta-function with the argument 3. For high momentum pions ($p^2 \simeq 4m^2$) the actual mass-dependence occurred for L_{st}:

$$L_{st} \simeq \left[\frac{e^{\sqrt{\mu^2+m^2}/T}}{3\,\alpha(N)\,m^2\,(\mu^2+m^2)^{3/4}\left(\frac{T}{2\,\pi}\right)^{3/2}\left(1 + \frac{15}{8}\frac{T}{\sqrt{\mu^2+m^2}}\right)}\right]^{\frac{1}{5}}, \tag{43}$$

at low T, and

$$L_{st} \simeq \left[\frac{\pi^2}{3\,\zeta(3)\,\alpha(N)\,m^2\,T^3}\right]^{\frac{1}{5}}, \tag{44}$$

at high temperatures, if $g^2\xi(m^2) << 1$, $\xi(m^2) \sim O(0.01/m^2)$ and $(\vec{p}^2/4m^2) << 1$ are valid in both temperature regime cases. Formula (42) reproduces the $\sim T^3$ behavior which is the same as the thermal distribution (in terms of density) for a gas of free relativistic massless particles. Such a behavior is expected anyway in high temperature limit if the particles can be considered as asymptotically free in that regime.

Actually, the increasing of T leads to squeezing of L_{st}, and $L_{st}(T = T_0) = R$ at some effective temperature T_0. The higher temperatures, $T > T_0$, satisfy to more squeezing effect and at the critical temperature T_c the scale $L_{st}(T = T_c)$ takes its minimal value. Obviously T_c defines the phase transition where the deconfinement will occur. Since all the masses tend to zero (chiral symmetry restoration) and $\alpha \to 0$ at $T > T_c$ one should expect the sharp expansion of the region with $L_{st}(T > T_c) \to \infty$. The following condition $\bar{n}(m,\beta)\cdot v_\pi = 1$ provides the phase transition (transition from hadronizing phase to deconfinement one) with the volume $v_\pi = (4\,\pi\,r_\pi^3/3)$, where r_π is the pion charge radius. Actually, the temperature of phase transition essentially depends on the charge (vector) radius of the pion which is a fundamental quantity in hadron physics. A recent review on r_π values is presented in.[15]

What we know about the source size estimation from experiments? DELPHI and L3 collaborations at LEP established that the correlation radius R decreases with transverse pion mass m_t as $R \simeq a + b/\sqrt{m_t}$ for all directions in the Longitudinal Center of Mass System (LCMS). ZEUS collaboration at HERA did not observe the essential difference between the values of R - parameter in $\pi^{\pm}\pi^{\pm}$, $K_s^0 K_s^0$ and $K^{\pm}K^{\pm}$ pairs, namely $R_{\pi\pi} = 0.666 \pm 0.009(stat) + 0.022 - 0.033(syst.)fm$, $R_{K_s K_s} = 0.61 \pm 0.08(stat) + 0.07 - 0.08(syst.)fm$ and $R_{K^{\pm}K^{\pm}} = 0.57 \pm 0.09(stat) + 0.15 - 0.06(syst.)fm$, respectively. The ZEUS data are in good agreement with the LEP for radius R. However, no evidence for $\sqrt{s}$ dependence of R is found. It is evidently that more experimental data are appreciated. However, the comparison between experiments is difficult mainly due to reference samples used and the Monte Carlo corrections.

Finally, our theoretical results first predict the L_{st} in (40) and (41), and both mass- and temperature - dependence are obtained clearly. This can serve as a good approximation to explain the LEP, Tevatron and ZEUS (HERA) experimental data. We need that the pion energies at the colliders are sufficient to carry these studies out (since the $\Delta\epsilon_p$ dependence). Careful simulation of their (pions) signal and background are needed. The more precisely measured pion momentum may be of some help. Also, determination of the final state interactions may clarify what is happening.

7. Conclusions

To summarize: we find the time dependence of the correlation function $C_2(Q)$ calculated in time-dependent external field provided by the operator $r(\vec{p}, t)$ and the chaotic coherence function $\alpha(m, \beta)$. Based on this approach we emphasize the explanation of the dynamic origin of the coherence in BEC, the origin of the specific shape of the correlation $C_2(Q)$ functions, and finding the dependence on the particle energy (and the mass) due to coherence function α, as seen from the QFT_β. Actually, the stochastic scale L_{st} decreases with the particle energy (the mass m). It is already confirmed by the data of LEP, Tevatron and HERA (ZEUS) with respect to the size of particle source.

In the framework of QFT_β the numerical analysis of experimental data can be carried out with a result where important parameters of $C_2{}^{--}(Q)$ and $C_2{}^{++}(Q)$ functions are retrieved (e.g., $C_0, R, \lambda, \epsilon, N, \alpha, L_{st}, T$).

The correlations of non-identical particles pairs can be observed and the corresponding C_2 parameters is retrieved. The off-correlation effect is given by the space-time distribution (14) containing the sum $\vec{p}_1 + \vec{p}_2$, and this effect is sufficient if the factor containing the sum $p_1^0 + p_2^0$ in (10) is not too small. The off-correlation effect is possible if the particle energies $p_i^0 (i = 1, 2)$ are small enough.

Besides the fact that like, e.g., $\pi^{\pm}\pi^{\pm}$ BEC the correlations $\pi^{\pm}\pi^{\mp}$ can serve as tools in the determination of parameters of the particle source. And besides the fact that these correlations play a particularly important role in the detection of random chaotic correction to BEC.

The stochastic scale L_{st} decreases with increasing temperatures slowly at low temperatures, and it decreases rather abruptly when the critical temperature is approached.

We claim that the experimental measuring of R (in fm) can provide the precise estimation of the effective temperature T_0 which is the main thermal character in the particle's pair emitter source (given by the effective dimension R) with the particle mass and its energy at given α fixed by $C_2(Q=0)$ and $\langle N \rangle$. Actually, T_0 is the true temperature in the region of multiparticle production with dimension $R = L_{st}$, because at this temperature it is exactly the creation of two particles occurred, and these particles obey the criterion of BEC.

We have found the squeezing of the particle source due to decreasing of the correlation radius R in the case of opposite charge particles. The off-correlated system of non-identical particles is less sensitive to the random force influence (α-dependence).

The results obtained in this paper can be compared with the static correlation function (see, e.g.,[16] and the references therein relevant to heavy ion collisions).

Finally, we should stress a new features of particle-antiparticle BEC which can emerge from the data. It is a highly rewarding task to experimental measurement of non-identical particles.

There is much to be done for $C_2(Q)$ investigation at hadron colliders. The time is ripe for dedicated searches for new effects in $C_2(Q)$ function at hadron colliders to discover, or rule out, in particular, the $\alpha(N)$ dependence.

In conclusion, the correlations of two bosons in 4-momentum space presented in this paper offer useful and instructive complimentary viewpoints to theoretical and experimental works in multiparticle femtoscopy and interferometry measurements at hadron colliders.

References

1. R. M. Weiner, *Phys. Rep.* **327**, 249 (2000).
2. R. Hanbury-Brown and R. Q. Twiss, *Philos. Mag.* **45** 633 (1954); *Nature* **177** 27 (1956); **178** 1046 (1956); **178** 1447 (1956).
3. G. A. Kozlov, O. V. Utyuzh and G. Wilk, *Phys. Rev.* **C68** 024901 (2003); G. A. Kozlov, *Phys. Rev. C* **58** 1188 (1998); *J. Math. Phys.* **42** 4749 (2001) and *New J. of Physics* **4** 23.1 (2002); G. A. Kozlov, *J. Elem. Part. Phys. Atom. Nucl.* **36** 108 (2005); G. A. Kozlov, L. Lovas, S. Tokar, Yu. A. Budagov and A. N. Sissakian, "Bose-Enstein correlations at LEP and Tevatron energies", hep-ph/0510027; G. A. Kozlov, O. Utyuzh, G. Wilk and Z. Wlodarczyk, "Some forgotten features of the Bose-Einstein correlations", hep-ph/0710.3710.
4. G. A. Kozlov, "BEC and the particle mass", hep-ph/0512184.
5. P. Langevin, *C. R. Acad. Sci.* **146** 530 (1908).
6. The ALEPH Collab., *Eur. Phys. J. C* **36** 147 (2004); The ALEPH Collab., *Phys. Lett. B* **611** 66 (2005).
7. The DELPHI Collab., *Phys. Lett. B* **379** 330 (1996); The DELPHI Collab., 2004-038 CONF-713.
8. G. Abbiendi et al. (OPAL Collab.), *Phys. Lett. B* **559** 131 (2003).
9. The ZEUS Collab., "Bose-Einstein correlations of neutral and charged kaons in deep

inelastic scattering at HERA", Report 276 at LP2005, Uppsala, Sweden 2005, *Phys. Lett. B* **583** 231 (2004).
10. T. Alexopoulos et al., [E735 Collaboration], *Phys. Rev. D* **48** 1931 (1993).
11. I. V. Andreev, M. Plumer and R. M. Weiner, *Phys. Rev. Lett.* **67** 3475 (1991).
12. G. Goldhaber et al., *Phys. Rev. Lett.* **3** 181 (1959); G. Goldhaber et al., *Phys. Rev.* **120** 300 (1960).
13. N. N. Bogolyubov, "Quasiaveragies in problems of statistical mechanics", JINR report D-781, JINR, Dubna (1961).
14. N. N. Bogolyubov and D. V. Shirkov, An Introduction to the Theory of Quantized Fields, (John Wiley and Sons, Inc.- Interscience, N.Y. 1959).
15. V. Bernard, N. Kaiser, and Ulf-G. Meissner, Phys. Rev. **C62** (2000) 028201.
16. C. Y. Wong and W. N. Zhang, *Phys. Rev. C* **76** 034905 (2007).

PION MULTIPLICITIES AT HIGH ENERGY PROTON NUCLEAR COLLISIONS IN FRAMES OF REGGEON ACTION

E. A. KURAEV[1] and S. BAKMAEV[2]

[1] *Joint Institute for Nuclear research, Dubna, Russian Federation,*
[2] *Bogolyubov Institute for Theoretical Physics, Kiev, Ukraine*

E. N. ANTONOV

Petersburg Nuclear Physics Institute, RU-188350 PNPI, Gatchina, St. Petersburg, Russian Federation

We study the pion production process at high energy proton-nuclei peripheral collisions. The dominant contribution for the case of non-exited final state hadrons arises from kinematics of central region of pionization. The basic mechanism consist in of creation of pair of real gluons by exchanged reggeized gluons with subsequent conversion to a pair of pions.

We show that the mechanisms which include emission of two gluons in the same effective vertex contribution dominate compared with one with the creation of two separate gluons. Numerical estimations of cross section of pair of charged pions production for LHC facility give the value or order $10mb$.

The problem of exceed of positively charged pions in cosmic ray interaction with earth atmosphere is as well considered. By close analogy to QED case of lepton pair creation at ion's collision we generalize our result to the case of pion pairs creation, providing to keep into account the screening effect.

The distribution of n pion pair production as a function of impact parameter is discussed. We estimate as well the behavior of total cross-section as a function of n as $\sigma_n \sim 1/(n^2 \cdot n!)$.

1. Introduction

The problem of unitarization of the BFKL Pomeron as a composite state of two interacting reggeized gluons[1] is actual problem of QCD[2] . In the circle of questions are the construction of three Pomeron vertices, and more complicated ones including as well the ordinary gluons can be solved using in terms of vertices of effective Regge action formulated in form of Feynman rules for effective Regge action build in[3] . In this paper the effective vertices containing the ordinary and the reggeized gluons were build.

A lot of attention to describe the creation in the peripheral kinematics of the bound states of heavy and light quarks was paid recent time[4] . Creation of gluon bound states (gluonium) in the peripheral kinematics was poorly considered in literature. This is a motivation of this paper.

To satisfy the requirement for the scattered nucleons to be colorless states (non excited barions), we must consider the Feynman amplitudes corresponding to two reggeized gluons exchange. More over in each nucleon both reggeized gluons must interact with the same quark.

Process of creation of two gluons with the subsequent conversion of them to the bound state with quantum number of σ meson in the pionization region of kinematics of nucleons collisions:

$$N_A(P_A) + N_B(P_B) \to N_A(P'_A) + N_B(P'_B) + g_1(p_1) + g_2(p_2);$$
$$g_1(p_1) + g_2(p_2) \to \sigma(p) \quad (1)$$

can be realized by two different mechanisms. One of them consist in creation of two gluons each of them is created in "collision" of two reggeized gluons RRP - type vertice. Another contains the effective vertex with two reggeized and two ordinary gluons RRPP - vertex. The second reggeized gluon in the scattering channel do not create real gluons. the relevant RRP vertex have a form:

$$V_\mu(r_1(q_1, a) + r_2(q_2, b) \to g(\mu, c)) = \Gamma^\mu_{cad}(q_1, k, q_2) = g f_{cad} C_\mu(q_1, k, q_2), \quad (2)$$

with the coupling constant g, $g^2 = 4\pi\alpha_s$ and $k = q_1 + q_2$,

$$C_\mu(q_1, k, q_2) = 2[(n^-)^\mu \left(q_1^+ + \frac{q_1^2}{q_2^-}\right) - (n^+)^\mu \left(q_2^- + \frac{q_2^2}{q_1^+}\right) + (q_2 - q_1)_\mu] \quad (3)$$

The 4-vector C_μ obey the gauge condition $k^\mu C_\mu = 0$.

The effective $RRPP$ vertex with the conservation law

$$r_1(q_1, c) + r_2(q_2; d) \to g_1(p_1, \nu_1, a_1) + g_2(p_2, \nu_2, a_2)$$

have a form[3] :

$$\frac{1}{ig^2}\Gamma^{\nu_1\nu_2}_{ca_1a_2d}(q_1, p_1, p_2; q_2) = \frac{T_1}{p_{12}^2} c^\eta(q_1, q_2)\gamma^{\nu_1\nu_2\eta}(-p_1, -p_2, k) +$$
$$\frac{T_3}{(p_2 - q_2)^2}\Gamma^{\eta\nu_1 -}(q_1, p_1 - q_1)\Gamma^{\eta\nu_2 +}(p_2 - q_2, q_2) -$$
$$\frac{T_2}{(p_1 - q_2)^2}\Gamma^{\eta\nu_2 -}(q_1, p_2 - q_1)\Gamma^{\eta\nu_1 +}(p_1 - q_2, q_2) - T_1[(n^-)^{\nu_1}(n^+)\nu_2 -$$
$$(n^-)^{\nu_2}(n^+)\nu_1] - T_2[2g^{\nu_1\nu_2} - (n^-)^{\nu_1}(n^+)\nu_2] - T_3[(n^-)^{\nu_2}(n^+)\nu_1 - 2g_{\nu_1\nu_2}] -$$
$$2q_2^2(n^+)^{\nu_1}(n_+)^{\nu_2}[\frac{T_3}{p_2^+ q_1^+} - \frac{T_2}{p_1^+ q_1^+}] -$$
$$2q_1^2(n^-)^{\nu_1}(n_-)^{\nu_2}[\frac{T_3}{p_1^- q_2^-} - \frac{T_2}{p_2^- q_2^-}] \quad (4)$$

with the light-like 4-vectors

$$n^+ = p_B/E, n^- = p_A/E, n^+n^- = 2, (n^\pm)^2 = 0$$

$\sqrt{s} = 2E$ is the total center of mass energy,

$$\gamma_{\mu\nu\lambda}(p_1,p_2,p_3) = (p_1-p_2)_\lambda g_{\mu\nu} + (p_2-p_3)_\mu g_{\nu\lambda} + (p_3-p_1)_\nu g_{\lambda\mu},$$
$$p_1+p_2+p_3 = 0 \qquad (5)$$

is the ordinary three gluon Yang-Mills vertex, and the induced vertices:

$$\Gamma^{\nu\nu'+}(q_1,q_2) =$$
$$= 2q_1^+ g^{\nu\nu'} - (n^+)^\nu (q_1-q_2)^{\nu'} - (n^+)^{\nu'}(q_1+2q_2)^\nu - \frac{q_2^2}{q_1^+}(n^+)^\nu (n^+)^{\nu'}; \quad (6)$$

$$\Gamma^{\nu\nu'-}(q_1,q_2) =$$
$$= 2q_2^- g^{\nu\nu'} + (n^-)^\nu (q_1-q_2)^{\nu'} + (n^-)^{\nu'}(-q_2-2q_1)^\nu - \frac{q_1^2}{q_2^-}(n^-)^\nu (n^-)^{\nu'} \quad (7)$$

We use here the notation $k^\pm = (n^\pm)_\mu k^\mu$ and light cone decomposition implies:

$$q_1 = q_{1\perp} + \frac{q_1^+}{2}n^-; q_2 = q_{2\perp} + \frac{q_2^-}{2}n^+; q_1^- = q_2^+ = 0,$$
$$p_i = \frac{p_i^+}{2}n^- + \frac{p_i^-}{2}n^+ + p_{i\perp}, p_\perp n^\pm = 0 \qquad (8)$$

The color structures are

$$T_1 = f_{a_1a_2r}f_{cdr}, T_2 = f_{a_2cr}f_{a_1dr}; T_3 = f_{ca_1r}f_{a_2dr} \qquad (9)$$

with f_{abc} is the structure constant of the color group; Jacoby identity provides the relation $T_1+T_2+T_3=0$. The conditions of Bose-symmetry and gauge invariance:

$$\Gamma^{\nu_1\nu_2}_{ca_1a_2d}(q_1,p_1,p_2,q_2)p_{1\nu_1} = 0, \Gamma^{\nu_1\nu_2}_{ca_1a_2d}(q_1,p_1,p_2,q_2) = \Gamma^{\nu_2\nu_1}_{ca_2a_1d}(q_1,p_1,p_2,q_2) \quad (10)$$

are satisfied.

2. Pomeron mechanisms of σ meson production

We consider the case when the hadrons after collision remains to be colorless. For the case of nucleon collisions it results that both exchanged reggeized gluons must interact with the same quark. The color coefficient associated with $RRPP$ vertex results to be

$$Trt^nt^c \times Trt^nt^d \times \Gamma^{\nu_1\nu_2}_{ca_1a_2d} = \frac{1}{4}N\delta_{a_1a_2}\Pi^{\nu_1\nu_2}_{02} \qquad (11)$$

$N=3$ is the rank of the color group. For the mechanism of creation of two separate gluons we have

$$Trt^nt^k \times Trt^mt^l \times f_{nma_1} \times f_{kla_2} \times \Pi^{\nu_1\nu_2}_{11} = \frac{1}{4}N\delta_{a_1a_2}\Pi^{\nu_1\nu_2}_{11} \qquad (12)$$

Projecting the two gluon state to the colorless and spin-less state we use the operator:

$$\mathcal{P} = \frac{\delta_{a_1 a_2}}{\sqrt{N^2-1}} \frac{g^{\nu_1 \nu_2}}{4} \tag{13}$$

The resulting expressions are

$$\frac{1}{16} N\sqrt{N^2-1}[\Pi_{02}, \Pi_{11}]$$

with

$$\Pi_{02} = -12 - [\frac{1}{(p_2-q_2)^2}\Gamma^{\eta\nu-}(q_1, p_1-q_1)\Gamma^{\eta\nu+}(p_2-q_2, q_2) + (p_1 \leftrightarrow p_2)] \tag{14}$$

$q_1 = l_1, q_2 = p_1 + p_2 - l_1$ and

$$\Pi_{11} = C_\mu(l-l_1, l_1-l+p_1)C_\mu(l_1, p_1-l_1), \tag{15}$$

Here l_1 is the 4-momentum of the gluonic loop, $l = P_A - P_{A'}$ is the transferred momentum.

In a realistic model describing interaction two reggeized gluons with the transversal momenta $\vec{l}_1, \vec{l}_2$, which form a Pomeron with quark[5] we have for the corresponding vertex

$$\Phi_P(\vec{l}_1, \vec{l}_2) = -\frac{12\pi^2}{N} F_P(\vec{l}_1, \vec{l}_2), \tag{16}$$

with

$$F_P(\vec{l}_1, \vec{l}_2) = \frac{-3\vec{l}_1\vec{l}_2 C^2}{(C^2 + (\vec{l}_1 + \vec{i}_2)^2)(C^2 + \vec{l}_1^2 + \vec{l}_2^2 - \vec{l}_1\vec{l}_2)} \tag{17}$$

and $C = m_\rho/2 \approx 400 MeV$. We note that this form of Pomeron-quark coupling obey gauge condition: it turns to zero at zero transverse momenta of gluons. The factor 3 corresponds to three possible choice of quark into proton.

Matrix elements of peripheral processes are proportional to s. To see it we can use Gribov's substitution to gluon Green functions nominators $g_{\mu\nu} = (2/s)P_{A\mu}P_{B\nu}$ with Lorentz index $\mu(\nu)$ is associated with $B(A)$ parts of Feynman amplitude. Performing the loop momenta l_1 integration it is convenient to use such a form of phase volume

$$d^4 l_1 = \frac{1}{2s} ds_1 ds_2 d^2\vec{l}_1, s_1 = 2P_A l_1, s_2 = 2P_B l_1 \tag{18}$$

Simplifying the nucleon nominators as

$$\bar{u}(P'_A)\hat{P}_B\hat{P}_A\hat{P}_B u(P_A) = s^2 N_A, N_A = \frac{1}{s}\bar{u}(P'_A)\hat{P}_B u(P_A), \sum |N_A|^2 = 2 \tag{19}$$

(with the similar expression for part B), we find all they be equal. The integration on variables $s_{1,2}$ can be performed for the sum of all four Feynman amplitudes as

$$\int_{-\infty}^{\infty} \frac{ds_{1,2}}{2\pi i}\left[\frac{1}{s_{1,2}+a_{1,2}+i0}+\frac{1}{-s_{1,2}+b_{1,2}+i0}\right]=1 \tag{20}$$

Combining all the factors the matrix element corresponding to the vertex RRPP can be written as

$$-iM = 2^5 3^2 s(\pi\alpha_s)^3 N_A N\sqrt{N^2-1}\frac{1}{N}F(\vec{\Delta})f(\vec{l},\vec{p}) \tag{21}$$

with $\vec{p}=\vec{p}_1+\vec{p}_2$ and the relative momenta of real gluons $\vec{\Delta}=(\vec{p}_2-\vec{p}_1)/2$,

$$f(\vec{l},\vec{p}) = \int \frac{d^2 l_1 C^4}{2\pi l_1^2(\vec{l}-\vec{l}_1)^2(\vec{p}-\vec{l}_1)^2} F_P(\vec{l}_1,\vec{l}-\vec{l}_1)F_P(\vec{l}_1-\vec{l},\vec{p}-\vec{l}_1)\Pi_{02}(\vec{l}_1,\vec{p}-\vec{l}_1) \tag{22}$$

and the similar expression for another mechanism. Here we had introduced the factor$F(\vec{\Delta}) = [a^2\vec{\Delta}^2+1]^{-2}$ which describe the conversion of two gluon state to bound state of the size $a \sim 1fm$ which is gluonium component of scalar meson.

Let perform the phase volume of the final state as

$$\begin{aligned} d\Gamma = \frac{d^3P'_B}{2E_{B'}}\frac{d^3P'_A}{2E_{A'}}\frac{d^3p_1}{2E_1}\frac{d^3p_2}{2E_2}(2\pi)^{-8}\delta^4(P_A+P_B-P_{A'}-P_{B'}-p_1-p_2) \\ = d\Gamma_{AB}d\Gamma_{12}(2\pi)^{-8}, \\ d\Gamma_{AB} = d^4l d^4P_{B'}d^4P_{A'}\delta^4(P_A-P_{A'}-l)\times \\ \times\delta^4(P_B+l-P_{B'}-p)\delta(P_{B'}^2-M_B^2)\delta(P_{A'}^2-M_A^2), \\ d\Gamma_{12} = \frac{d^3p}{2E_1}\frac{d^3\Delta}{2E_2} \end{aligned} \tag{23}$$

Using the relation

$$d^4l = \frac{1}{2s}d(2lP_B)d(2lP_A)d^2\vec{l} \tag{24}$$

we perform integration on the momenta of the scattered nucleons with the result:

$$d\Gamma_{AB} = \frac{d^2\vec{l}}{2s} \tag{25}$$

Keeping in mind the almost collinearity of 3-momenta of real gluon we can transform the gluon part of the phase volume as:

$$d\Gamma_{12} = d^4p\delta(p^2-M_\sigma^2)\frac{2}{M_\sigma}d^3\Delta \tag{26}$$

Integration on the $\vec{\Delta}$ can be performed in the explicit form:

$$\int d^3\Delta F^2(\Delta) = \frac{\pi^2}{4a^3} = \frac{2\pi^2 M_p^3}{10^3 a(fm)^3} \tag{27}$$

where we had used the conversion constant $M_p \times fm = 5$, M_p is the nucleon mass. The last part of the phase volume of gluonic system can be arranged using light cone form of 4-momentum p (see (8)):

$$d^4p\delta(p^2 - M_\sigma^2) = \frac{dp^+}{2p^+}d^2\vec{p} = \frac{1}{2}Ld^2\vec{p} \tag{28}$$

with the so called "boost logarithm" $L \approx \ln(\frac{2E}{M})$, $E, M-$ Energy and mass of proton in laboratory reference frame. For LHC facility as well as for cosmic protons (in the knee region of spectra) we use below $L = 15$.

For the contribution to the total cross section we obtain

$$\sigma_1^P = A\frac{\alpha_s^6 LM_p^3}{M_\sigma M_\rho^4}J, A = \frac{6^4}{5^3}\frac{N^2-1}{N^2}\frac{\pi^2}{a(fm)^3} \tag{29}$$

with

$$J = \int \frac{d^2l d^2p}{(2\pi)^2 C^4} f^2(\vec{l}, \vec{p}) \tag{30}$$

Numerical integration give $J = 7.4 * 10^3$.

Corresponding contribution to the total cross section of the single σ meson production is of order of $10mb$.

The contribution arising from the other mechanism of production (including the interference of amplitudes) turns out be at least order of magnitude less. It determines the accuracy of the result obtained on the level of 10%.

3. Screening effects. Several σ- production

Let us now generalize the result to include the screening effects as well as the possibility to produce several σ mesons.

At large impact parameters limit proton interact with the whole gluon field of the nucleon (or nuclei) moving in the opposite direction coherently. So the main contribution arises from the many Pomeron exchanges mechanism (compare with the "chain' mechanism essential in BFKL equation)[1] . So we must consider Pomeron s-channel iterations. Let consider three kinds of the iteration blocks. One is the pure Pomeron exchange, the second is Pomeron with the sigma meson emission from the central region. The third one is the "screening block":two blocks of the second type with the common virtual σ-meson. Contribution to the amplitude of production n σ mesons of the blocks of the third kind is associated with the "large logarithm" which arises from boost freedom of these blocks in completely analogy with QED[6] .

In the similar way the closed expression (omitting the terms of order $1/N^2$ compared with the ones of order of 1) for the summed on number of s-channel iteration ladders of the first and the third type can be obtained using the relation

$$\int \Pi_1^n \frac{d^2k_i}{(2\pi)^2} = \int \Pi_1^{n+1}\frac{d^2k_i}{(2\pi)^2}\int d^2\rho \exp(i\vec{\rho}\sum_1^{n+1}(\vec{k}_i - \vec{q})) =$$
$$\int d^2\rho \exp(-i\vec{q}\vec{\rho})\Pi_1^{n+1}\frac{d^2k_i}{(2\pi)^2}e^{i\vec{k}_i\vec{\rho}} \tag{31}$$

Accepting the assumption about color-less structure of the Pomeron as a bound state of two reggeized interacting gluons and applying the same sequence of transformations as was done in[6] we obtain for the cross section of n σ mesons production at the peripheral high energy protons collisions:

$$\sigma_n = \int \frac{d^2\rho}{(2\pi)^2} \frac{(L\sigma_0 Z(\rho))^n}{n!} exp(-L\sigma_0 Z(\rho)) \tag{32}$$

with $\sigma_0 = 2.25 * 10^{-3}\alpha_s^6 M_p^3/(M_\sigma M_\rho^4 a(fm)^3)$ and

$$Z(\rho) = \int \frac{d^2p}{(2\pi)^2}|B(\rho,p)|^2,$$
$$B(\rho,p) = \int \frac{d^2l}{C^2} f(l,p) exp(i\vec{l}\vec{\rho}) \tag{33}$$

Numerical estimations of the cross section of one sigma meson production for the give $\sigma_1 = 10mb$.

Distribution in impact parameter can be written down as:

$$\frac{d\sigma_n}{d^2\rho} = \frac{1}{(2\pi)^2} \frac{(L\sigma_0 Z(\rho))^n}{n!} exp(-L\sigma_0 Z(\rho))$$

.

Keeping in mind the exponential decrease of $Z(\rho)$ at large ρ: $Z(\rho) \sim c_1 e^{-\frac{-C\rho}{2}}$, we can estimate the behavior of σ_n at large n; $\sigma_n \sim C^{-2} \cdot \frac{1}{n^2 n!}, n \gg 1$.

4. Conclusion

Reggeized gluons exchange iteration in $s-$ channel dominated being compared with t-channel iteration (critical Pomeron formation), because in the first case effective expansion parameter is $\alpha_s \cdot A$ where A is number of nucleons in the nuclear. We remind that in $t-$channel it is α_s.

The contribution of enhanced Pomeron exchange (included three Pomeron vertices) can be estimated of order 10% of result obtained here. These reasons determine the level of accuracy our results.

The paper[?] concerns one real gluon production, at three reggezeid gluon collision. This mechanism implied production colored states of gluon and correspond exitation nuclei. Which is not a subject of our consideration.

The results given above can be applied to explain the excess of the positive muons compared with the negative ones produced by cosmic ray interaction with the Earth surface. Really one can neglect QED mechanisms of production of the charged pions in favor to strong interactions one. It turns out that the main mechanism is the peripheral production of the pions pairs (with the subsequent decay to muons) in the high energy cosmic ray proton collisions with the nuclei of nitrogen or oxygen in the Earth atmosphere.

Keeping in mind atmosphere density about $10^{19}cm^{-3}$ with thickness $\sim 10km$, effective proton size $\sim 1.4fm$, we obtain number of effective collision $N_{eff} = (1.4fm)^2 \cdot 10^5 cm \cdot 10^{19} cm^{-3} \sim 1$.

QED mechanism contribution is suppressed by at least by factor $(\frac{\alpha}{\alpha_s})^4 < 10^{-3}$. Contribution of peripheral interaction of proton with gas nuclei with impact parameter $\rho \gg A^{1/3} fm$ is small. For a $\rho > 10fm$ suppress by factor $e^{m_\rho \cdot \rho/2} < 10^{-5}$ appears. So the only mechanism of penetration of proton throw nuclear becomes dominant.

Travelling through the nuclei the cosmic proton have a direct collision with the protons and the neutrons of the nuclei $\rho \leq 1fm$ (such kind of collisions produce positive charged pions due to the decay of the excited resonances) or have the peripheral collisions, when the pairs of pions are produced ($1fm < \rho < A^{\frac{1}{3}} fm$).

The number of positive charged pions produced in direct collisions N_d, is proportional to $A^{1/3}$ with the characteristic atomic number $A = 14$. The number of pion pairs produced in the peripheral collisions can be estimated as

$$N_p = M_p^2 \sigma_1 \approx 7.8 \tag{34}$$

For the ratio of the positive charged muons to the negative charged ones $R = \frac{N_{\mu+}}{N_{\mu-}}$ we have

$$R_{th} = 1 + \frac{N_d}{N_p} = 1.32 \tag{35}$$

This quantity can be compared with the recent experimental value[7] (here only hard muons are taken into account).

$$R_{exp} = \frac{N_{\mu^+}}{N_{\mu_-}} = 1.4 \pm 0.003 \tag{36}$$

These values are in reasonable agreement.

Reggeized gluons exchange iteration in $s-$ channel dominated being compared with t-channel iteration (critical Pomeron formation), because in the first case effective expansion parameter is $\alpha_s \cdot A$ where A is number of nucleons in the nuclear. Whereas for $t-$channel it is α_s.

The contribution of enhanced Pomeron exchange (included three Pomeron vertices) can be estimated of order 10% of result obtained here. These reasons determine the level of accuracy our results.

The large number n of pair produced in central region σ_n falls with n more fast than factorial $\sigma_n \sim \frac{\sigma_1}{n^2 n!}$

Acknowledgement

We are grateful to E. Kokoulina for taking part at the initial state of this work. We are grateful to A. Kotinjan and O. Kuznetsov for discussing the possibility to study the Pomeron physics at COMPASS facility with pion beams, E. Tomasi-Gustafsson and R. Peschanskii for discussions. We are grateful to a grants INTAS 05-1000-008-8323; MK-2952.2006.2 for financial support.

References

1. E. Kuraev, L. Lipatov and V. Fadin,*Sov. Phys.JETP*, **44**, 443 (1976); ibid **45**, 199 (1977); I. Balitski and L. Lipatov, *Sov. J. Nucl. Phys.* **28**, 822 (1978).
2. V. Gribov and L. Lipatov, *Sov. J. Nucl. Phys.* **18**, 438,675 (1972);
   L. Lipatov, *Sov. J. Nucl. Phys.* **20**, 93 (1975); G. Altarelli and G. Parisi, *Nucl. Phys. B* **126**, 298 (1977); Yu. Dokshitzer, *Sov.Phys. JETP* **46**, 641 (1977).
3. E. N. Antonov, L. N. Lipatov, E. A. Kuraev and I. O. Cherednikov, *Nucl. Phys. B* **721**, 111 (2005); [arXiv:hep-ph/0411185].
4. V. Kiselev, F. Likhoded, O. Pakhomova and V. Saleev, *Phys. Rev. D* **66**, 034030 (2002).
5. J. Gunion and D. Soper, *Phys. Rev. D* **15**, 2617 (1977);
   M. Fukugita and J. Kwiecinski, *Phys. Lett. B* **83**, 119 (1979).
6. E. Bartos, S. Gevorkyan, E. Kuraev and N. Nikolaev, *Phys. Lett. B* **538**, 45 (2002).
7. P. Adamson *et al.* [MINOS Collaboration], *Phys. Rev. D* **75**, 092003 (2007); [arXiv:hep-ex/0701045].

R_{BEC} ANALYSIS AND THE CASCADE MODEL

V. ROINISHVILI

Joint Institute for Nuclear Research,
Joliot Curie st., 6, Dubna, Moscow region, 141980 Russia;
Andronikashvili Institute of Physics
Tamarashvili st. 6, Tbilisi, Georgia
E-mail: Vladimir.Roinishvili@ihep.ru

A simple cascade model which relates the energy dependence of mean charge multiplicity to the growth of the total cross-sections of strong interactions is presented. The main assumption, that the total time of hadronization is limited by the characteristic time of the strong interaction is explicitly used. The model describes well the available data on e^+e^- annihilation and $pp/\bar{p}p$ and is in qualitative agreement with the main results of the Bose-Einstein Correlation analysis in the hadron physics.

1. Introduction

The proposed model of multiparticle production is based on the statement that the total time needed for all successive in time steps of hadronization must be equal to the characteristic time of the strong interaction. It cannot be larger by definition and cannot be much smaller since in this case an additional step of hadron production has a chance to happen. Such approach leads to the condition

$$N_{step} \cdot \langle \tau \rangle = \tau_{\rm s}, \tag{1}$$

where N_{step} is the number of the steps of the hadronization process, $\tau_{\rm s}$ is the characteristic time of the strong interaction and the factor $\langle \tau \rangle$, which has the dimension of time, may be considered as a mean time elapsed between two consecutive steps. In the space coordinates $\tau_{\rm s}$ is the characteristic size of the range of strong interactions and $\langle \tau \rangle$ is the mean distance between two steps. Both $\langle \tau \rangle$ and $\tau_{\rm s}$ are considered in the same frame, which could be the c.m.s. for colliders or the rest frame of heavy particles decaying into hadrons.

It is obvious that creation of a large number of particles simultaneously is very improbable. Due to the conservation laws, each created particle before leaving the range of strong interaction has to interact somehow with all the others to share properly the available energy, momentum etc. In this case the number of sequential interactions is proportional to at least the first power of the number of produced particles, and condition (1) suppresses high multiplicity events very strongly.

More economical are consecutive and parallel in time (or space) ways of hadronization of energy, known as cascade models. A version of the cascade model which satisfies condition (1) is presented below.

2. Cascade Model (CM) of multiparticle production ([1])

In accordance with the cascade models (see for example[2]) we suppose that the condensed energy fragments into two objects (resonances, clusters, strings...) while conserving energy, charge, etc. via creating, for example, pairs of quarks. Each object then fragments independently again into two parts and so on until the stable hadrons are formed. (Note that in[2] the number of preceding steps of a created particle is called the "age" of the particle.) In the case of charged particle production N_{step} in Eq.(1) is the mean number of successive in time steps (or the mean "age" charge particles) of the cascade process in which new pairs of charged objects or particles are created. Then

$$n_{\rm ch} = 2^{N_{\rm step}} = 2^{\frac{\tau_s}{\langle\tau\rangle}}, \tag{2}$$

where $n_{\rm ch}$ is the mean number of either directly produced charged hadrons or the charged hadrons originating from the strong decays of the objects with lifetime not exceeding $\tau_{\rm s}$.

The value of $\tau_{\rm s}$ must be of the order of m_π^{-1} since pions are the lightest hadrons. However, the cross-sections of strong interactions grow with energy. According to the parameterization used in[3] , the rise of the total cross-sections at high energies, for all interactions in which secondary hadrons are produced, is described by the term

$$\sigma_{\rm r}(s) = B \cdot \log^2(s/s_0), \tag{3}$$

valid in the range of $s > s_0$.

It is important to note, that the values of parameters $B = 0.308 \pm 0.01$ mb and $\sqrt{s_0} = 5.38 \pm 0.50 GeV$ are common for all interactions: pp, $\bar{p}p$, πp, Kp, γp, $\gamma\gamma$... This means that the rise of the cross-sections with energy is not a property of the colliding particles, but rather the property of strong interactions. The radius of the interaction range, $r_{\rm s}$, is proportional to the square root of the cross-section and thus $\tau_{\rm s}$ must depend on energy. We assume

$$\begin{aligned} \tau_{\rm s} &= \frac{1}{m_\pi} + \frac{1}{\hbar c}\sqrt{\sigma_{\rm r}(s)} \\ &= \tau_s^o \cdot \left[1 + \sqrt{\frac{\sigma_{\rm r}(s)}{\sigma^0}}\right], \end{aligned} \tag{4}$$

where $\tau_{\rm s}^o = 1/m_\pi$, $\sigma^0 = (\hbar c/m_\pi)^2$ and $\sqrt{s}$ is the energy available for hadronization: the total energy in annihilation processes, the mass of a heavy particle decaying into hadrons, or the fraction of energy accessible for particle productions in pp, $\bar{p}p$, πp... interactions.

Eq. (2) can be easily transformed into the form

$$n_{ch} = A \cdot \left(\frac{s}{s_0} \right)^{\alpha} \tag{5}$$

where $A = 2^{\frac{\tau_s^O}{\langle \tau \rangle}}$ and $\alpha = \log(A) \cdot \sqrt{\frac{B}{\sigma^0}}$. Such power-like energy dependence of the mean multiplicity results from many different approaches, all of which have at least two free parameters.

3. Comparison of the CM with the experimental data

3.1. $e^+e^- \to h^\pm + X$

For e^+e^- annihilation, the data extracted from COMPAS database system[4] were used with $\sqrt{s} > 10$ GeV, in order to be far from the energy limit s_0 for validity of Eq.(3,4). Statistical and systematic errors were added in quadrature. Most of the data points include charged particles from K^0 and Λ decays. Theories or models of strong interactions should predict the multiplicity of particles produced via strong interactions only, therefore the data were corrected for these decays using the measured yield of K^0 and Λ at various energies[5] . This correction amounted to about 10%.

In Fig.1 the experimental data for the mean charged multiplicity (hollow circles) are shown together with the results of the fit using Eq.(2) with one free parameter $\langle \tau \rangle^{-1}$. The quality of the fit is very good, with $\chi^2/\text{n}.d.f. = 24.4/39$ for the obtained value of $\langle \tau \rangle^{-1} = 0.3455 \pm 0.0003(st) \pm 0.0057(syst)$ GeV where the systematic error includes uncertainties in B and s_0. Also plotted are the data points for Z[6] and W[7] decays (corrected for K^0 and Λ decays, as above): $Z \to b\bar{b}$ (the black square), $Z \to h^\pm + X$ (the triangle), $Z \to u\bar{u}, d\bar{d}...$ (the hollow square) and $W \to h^\pm + X$ (the black circle). The mean multiplicities for the first two decays are higher than the values predicted by CM, because they include a large fraction of the particles from weak decays of b quarks (the branching ratio of $Z \to b\bar{b}$ is about 15%). The measured values of the mean multiplicity for $Z \to u\bar{u}, d\bar{d}...$ and W decays are in a very good agreement with CM. This indicates that the process of hadronization is the same for both e^+e^- annihilation and the decays of heavy particles.

Since $\langle \tau \rangle$ is the only unknown parameter in the model, it can be calculated for each data point. The results are shown in the left top corner of Fig.1. One can see that the value of $\langle \tau \rangle$ is stable in the full available energy range.

3.2. $pp/\bar{p}p \to h^\pm + X$

3.2.1. *pp and $\bar{p}p$ data.*

The highest energy data ($\sqrt{s} > 100$ *GeV*) available for the mean multiplicity of charged particles is from $\bar{p}p$ collisions at UA5[8] which are for non-single-diffractive events (NSDE) and do not include particles from K^0 and Λ decays. At lower energies, in order to have a uniform sample, we used the data on NSDE in pp interactions at ISR[9] which also do not include particles from weak decays.

3.2.2. *Inelasticity*

For the pp and $\bar{p}p$ interactions the situation is more complicated than for e^+e^- because of specific dynamics: there are quasi-elastic interactions and diffractive processes producing a small number of hadrons but carrying away the bulk of the total energy. The lump of energy that hadronizes is not at rest in the centre-of-mass system. Therefore only a fraction of energy

$$K \cdot \sqrt{s} \tag{6}$$

is available to create new particles where the so-called inelasticity K may itself depend on energy.

Let us define the inelasticity K_{ch} as:

$$K_{\text{ch}} = \frac{s^{1/2}_{e^+e^-}(n_{\text{ch}})}{s^{1/2}_{pp/\bar{p}p}(n_{\text{ch}})} \tag{7}$$

where $s^{1/2}_{pp/\bar{p}p}(n_{\text{ch}})$ is the energy of the hadron colliders at which mean charge multiplicity is equal to n_{ch}, while $s^{1/2}_{e^+e^-}(n_{\text{ch}})$ is the energy of e^+e^- annihilation corresponding to the same multiplicity.

On the top left corner of Fig.2 K_{ch} is plotted versus $\sqrt{s}$ for the pp interactions. These points refer to the multiplicity range where the data is available for both e^+e^- and pp collisions. It is clear that for NSDE the inelasticity decreases with energy. The curve is the result of the fit using the expression:

$$K_{\text{ch}} = K_0 \times \left(\frac{s}{s_0}\right)^{-\epsilon} \tag{8}$$

with obtained parameter values:

$$K_0 = 1.06 \pm 0.20 \ \ and \ \ \epsilon = 0.081 \pm 0.044. \tag{9}$$

Note that the value of ϵ is, within its error, inside the range $\epsilon \leq 0.06$ mentioned in[10] .

The inelasticity as defined in (7) cannot be measured directly for $\bar{p}p$ because there is no data in the corresponding range of multiplicity in e^+e^- annihilation.

3.2.3. *pp and $\bar{p}p$ multiplicity*

In Fig.2 the experimental data on mean charge multiplicity in pp (black circles, ISR) and $\bar{p}p$ (hollow circles, UA5) are plotted. The curve represents Eq.(2), with $\langle\tau\rangle$ fixed to the value $\langle\tau\rangle^{-1} = 0.345$ GeV obtained from e^+e^- data, and s replaced by $K^2_{\text{ch}} \cdot s$ with the obtained values of parameters (9). The good agreement of the $\bar{p}p$ data (not used for K_{ch} estimation !) with prediction of CM indicates that energy dependence of K_{ch} is the same for both pp and $\bar{p}p$. The $\chi^2/n.d.f. = 2.35/3$. The data points shown as squares, related to the Fermilab energies, are taken from[11] . They were not used in the χ^2 calculation because in[11] there is no mention of the associated errors. However, one can see that these points are also in a in good agreement with the model, even if the errors are small.

If $K_{\rm ch}$ continues decrease as (8) up to 14 TeV, then the predicted value of the mean multiplicity for the LHC is $\langle n_{ch} \rangle = 96 \pm 5(syst)$.

4. CM and R_{BEC} results

It is known, that the Bose-Einstein Correlation analysis in the hadron physics were initiated by the results of the astronomy physics on measurements of the size of stars. The astronomers define the size of stars by measurements of the energy difference between two photons emitted from the same star. Since photons follow Bose-Einstein statistic they observe a sharp peak at the small differences. The width of the peak directly correlates with the radius of the star.

In the hadron sector physicits measure Δp - the 4-momenta difference between two like-sign mesons and observe a sharp peak at small Δp also. From the width of the peak they define the radius of the hadronization region.

But there are some essential differences between the astronomy and the hadron physics:

- Astronomy: the sources of photons - atoms, are distributed over the surface of the star. The photons from the inside of the star are absorbed.
  Hadron: the sources of stable hadrons - resonances ($\rho, \omega, ..$) may be distributed over the full volume of the hadronization region.
- Astronomy: the sources of photons are at the rest in the rest system of the star.
  Hadron: the sources of stable hadrons have a rather high momentum in the rest system of the hadronization region.
- Astronomy: the size of the star is constant.
  Hadron: the size of the hadronization region may fluctuate.

Having in mind these differences it is very difficult to make unambiguous conclusions about the radius of hadronization - R_{BEC} measured by BEC in different experiments. The measured values of R_{BEC} differ from one experiment to another, depending on the conditions of experiments - trigger, event selections, the data analysis procedure. But there are two rather well established properties of R_{BEC}:

1. R_{BEC} is almost energy independent.
2. R_{BEC} increases with multiplicity.

These two properties of R_{BEC} can be explained by the CM.

1. It is very improbable to find two pions with the same 4-momenta from different branchings if multiplicity is low (the case of the mean multiplicity, which is much less then avaiable one). This is because the sources of pions (ρ, ω, ...) have rather high momenta and they are moving in different directions. But it is possible to find them from the same branching and with quite high probability issued from the neighbouring steps. Then R_{BEC} measures just the mean distance between two steps

- $\langle \tau \rangle$ which is energy independent, as it was shown above.

2. If one selects events with a very high multiplicity (near to the maximum available) when sources of pions (ρ, ω, ...) are almost at the rest in the rest system of the hadronization region and are distributed over the surface, then the probability to detect the pions with the same momenta from the different branchings becomes large. In this case the situation is similar to the astronomy and R_{BEC} measures the radius of the full region of hadronization. Fig.3 shows the R_{BEC} dependency on $\log(n_{ch})$ for different data analysis procedures[12,13] . Although the measured values of R_{BEC} cannot be compare directly for this two experiments (because of different experimental conditions) both of them show a strong dependence of R_{BEC} on the multplicity. One should mention, that a linear dependency of R_{BEC} on n_{ch} is not excluded due to the errors.

Thus the CM explains why R_{BEC} is almost energy independent, but strongly depends on the mean number of steps - $N_{step} \backsim \log(n_{ch})$.

5. Conclusions

The presented version of the Cascade Model:

- describes very well the dependency of mean multiplicity on $\sqrt{s}$ over all available energies with $\sqrt{s} > 10\ GeV$;
- shows that the mean time (or distance) elapsed between any two consecutive steps of the cascade process is energy independent in the available range of energies and is equal to $(0.345\ \mathrm{GeV})^{-1}$ (or about $0.6 fm$);
- describes qualitatively the behavior of R_{BEC} - the energy independence but strong dependence on the multiplicity.

The measured inelasticity K_{ch} as defined in (7) for non-single-diffractive events in pp interactions decreases with energy. This can be an indication that $pp/\bar{p}p$ interactions became more peripheral with the increase of energy.

Acknowledgments

The author is grateful to A.Volodko, J.Mandjavidze and many other physicists for very useful discussions and to A.Sisakian for his interest to this work and his support.

References

1. V.Roinishvili, Pr IHEP 2004-31, 2004
2. A.M.Polyakov, *JETP* **5**, issue 5, 1572 (1971)(in russian)
3. a) J. R. Cudell *et al.*, *Phys. Rev. D* **65**, 074024 (2002), [arXiv:hep-ph/0107219];
   b) K. Hagiwara *et al.* [Particle Data Group Collaboration], *Phys. Rev. D* **66**, 010001 (2002).

4. http://wwwppds.ihep.su:8001/ppds.html S. Eidelman *et al.* [Particle Data Group Collaboration], *Phys. Lett. B* **592** 1 (2004).
5. Ref.[3b] p.256.
6. P.Abreu et al., *Eur.Phys.J. C* **6**, 19 (1994).
7. Ref.[3b] p.285.
8. R.E.Ansorge et al., *Z.Phys. C* bf 43, 357 (1989); G.J.Alner et al.,*PL B* **160**, 193 (1985).
9. A.Breakstone et al.,*Phys.Rev. D* **30**, 528 (1984).
10. Z.Wlodarczyk, *J. Phys. G: Nucl.Part.Phys.* **21**, 281 (1995).
11. F.S.Navarra et al., *Phy.Rev. D* **67**, 114002-1 (2003)
12. C.Albajar et al., *Phys.Lett. B* **226**, 410 (1989).
13. T.Alexopoulos et al., *Phys.Rev. D* **48**, 1931 (1993).

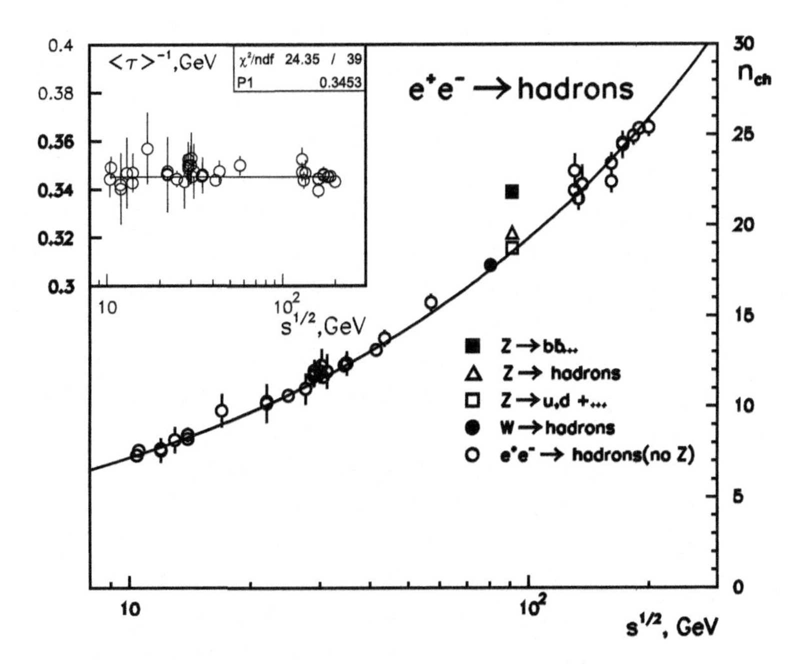

Fig. 1. Energy dependence of the mean charge multiplicity for e^+e^- annihilation and the data on hadronic decays of Z and W. The curve represents the result of the fit to $e^+e^- \to$ hadrons only (see text). On the left top corner $\langle\tau\rangle^{-1}$ is plotted vs energy.

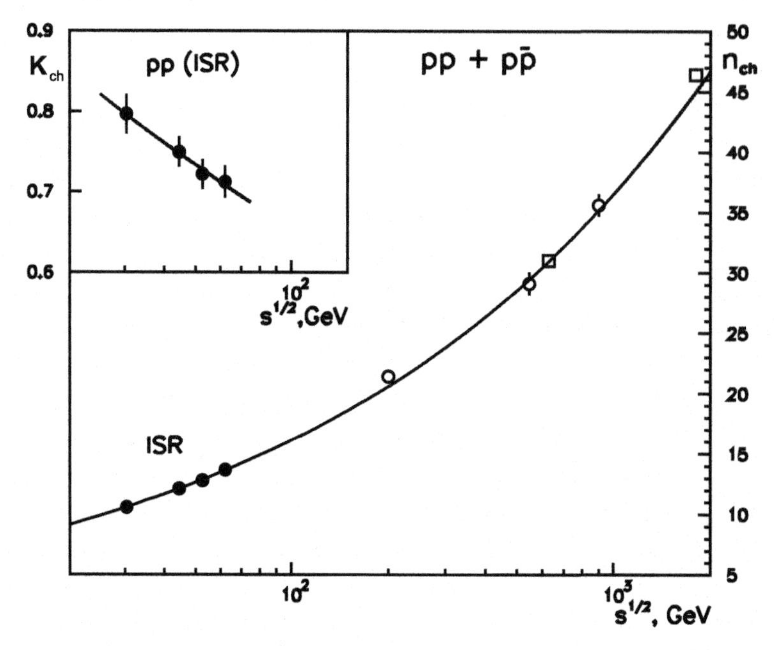

Fig. 2. Energy dependence of the mean charge multiplicity for $pp/\bar{p}p$ collisions. Circles represent the ISR and UA5 data, squares — Fermilab data from[11] . The curve shows the expectation of CM. On the top left corner, the inelasticity $K_{\rm ch}$ is shown vs the energy of pp interactions. The curve is the result of the fit (see text). Note that only ISR data (black circles) were used in the fit.

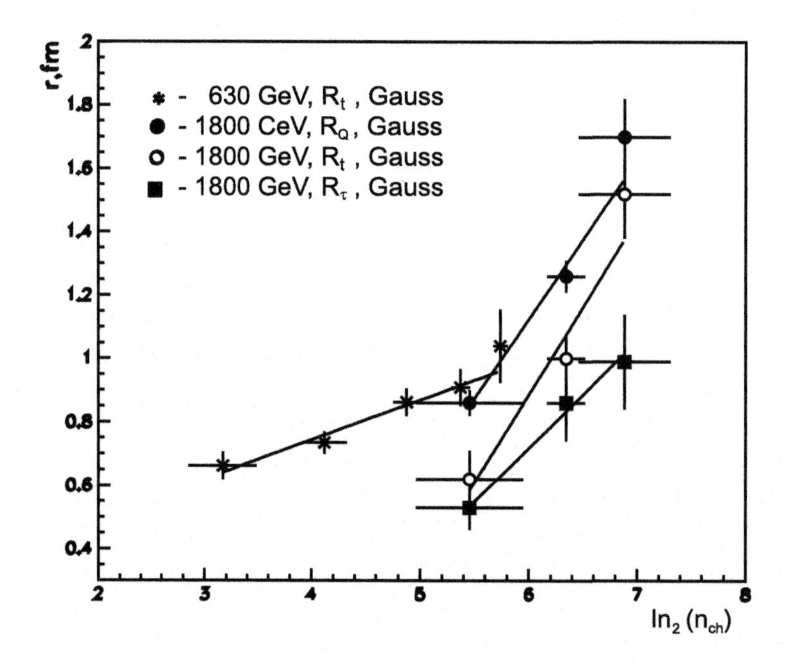

Fig. 3. R_{BEC} as a function of $\log(n_{ch})$.

J/Ψ ELECTROPRODUCTION IN A DUAL MODEL

R. FIORE[1] , L. L. JENKOVSZKY[2] , V. K. MAGAS[3] , A. PROKUDIN[4]

[1] *Universita' della Calabria, Dipt. di Fisica*
Arcavacata di Rende, I-87030 Cosenza, Calabria, Italy,
[2] *Bogolyubov Institute for Theoretical Physics (BITP),*
Ukrainian National Academy of Sciences
MKI, KF14-b, Metrolohichna str., Kiev, 03143, Ukraine, and
RMKI, KFKI, Budapest, Hungary,
[3] *Departament d'Estructura i Constituents de la Matéria,*
Universitat de Barcelona, Diagonal 647, 08028 Barcelona, Spain
[4] *Dipartimento di Fisica Teorica, Università di Torino and*
INFN, Sezione di Torino, Via P. Giuria 1, I-10125 Torino, and
Di.S.T.A., Università del Piemonte Orientale "A. Avogadro", Alessandria, Italy

J/Ψ photo- and electroproduction is studied in the framework of the analytic $S-$matrix theory. It is argued that at low energies the background, which is the low-energy equivalent of the high-energy diffraction, replaces the Pomeron exchange. Detailed calculations and comparison with the data for the on shell J/Ψ photoproduction have been done in[1] , based on Dual Model with Mandelstam Analyticity (DAMA). Here we discuss the generalization of the model for J/Ψ electroproduction in the framework of the off mass shell extension of DAMA[2] .

1. Introduction

J/Ψ photo- and electroproduction is a unique testing field for diffraction[1] . Most of the theoretical approaches to J/Ψ photoproduction are based on the Pomeron or multi-gluon exchanges in the t-channel of the reaction (for a review see[3]). A common feature of these models is the uncertainty of the low-energy extrapolation of the high-energy exchange mechanisms. The missing piece is the low-energy background contribution, significant between the threshold and the region of the dominance of the exchange mechanism, e.g., the onset of the Regge-Pomeron asymptotic behaviour.

What is the fate of the Regge exchange contribution when extrapolated to low energies? The answer to this question was given in late 60-ies by dual models: the proper sum of direct-channel resonances produces Regge asymptotic behavior and vice versa. Thus to take the sum of the two (the so-called interference model) is incorrect, it will lead to the double counting.

While the realization of the Regge-resonance duality was quantified within narrow-resonance dual models, notably in the Veneziano model[4] , a similar solution for the Pomeron (=diffraction) was not possible in the framework of that model, just because of its narrow-resonance nature. Our poor understanding of the "low-energy diffraction", or the background, is caused also by difficulties in its separation (discrimination and identification) from the resonance part.

Experimentally, the identification of these two components meets difficulties coming from the flexibility of their parameterizations. The separation of a Breit-Wigner resonance from the background, as well as the discrimination of some $t-$channel exchanges with identical flavor content and C-parity is a familiar problem in the experimental physics.

Unique reactions are J/Ψ photo- and electroproduction; it is purely diffractive since resonances here are not produced and, due to the Okubo-Zweig-Iizika (OZI) rule[5] , no flavor (valence quarks) can be exchanged in the t-channel. Thus this reaction is an ideal opportunity to study diffraction[1] , as we shall see below.

Dual models with Mandelstam analyticity (DAMA)[6] appeared as a generalization of narrow-resonance (e.g. Veneziano) dual models, intended to overcome the manifestly non-unitarity of the latter. Contrary to narrow-resonance dual models, DAMA does not only allows for, but moreover requires non-linear, complex trajectories. This property allows for the presence in DAMA of finite-width resonances and a non-vanishing imaginary part of the amplitude. The maximal number of direct resonances is correlated with the maximal value of the real part of the relevant trajectory. An extreme case, when the real part of the trajectory terminates below the spin of the lowest ($s-$channel) resonance, is called "super-broad-resonance approximation"[7] . The resulting scattering amplitude can describe the non-resonating direct-channel background, dual to the Pomeron exchange in the $t-$channel. The dual properties of this construction were studied in.[8]

The (s, t) term (s and t are Mandelstam variables) of a dual amplitude with Mandelstam analyticity (DAMA)[6] is given by:

$$D(s,t) = c\int_0^1 dz \left(\frac{z}{g}\right)^{-\alpha(s')-1} \left(\frac{1-z}{g}\right)^{-\alpha_t(t'')}, \tag{1}$$

where $\alpha(s)$ and $\alpha(t)$ are Regge trajectories in the s and t channel correspondingly; $x' = x(1-z)$, $x'' = xz$ $(x = s, t, u)$; g and c are parameters, $g > 1$, $c > 0$.

For $s \to \infty$ and fixed t DAMA is Regge-behaved $D(s,t) \sim s^{\alpha_t(t)-1}$. The pole structure of DAMA is similar to that of the Veneziano model except that multiple poles appear on daughter levels[6] .

$$D(s,t) = \sum_{n=0}^{\infty} g^{n+1} \sum_{l=0}^{n} \frac{[-s\alpha'(s)]^l C_{n-l}(t)}{[n-\alpha(s)]^{l+1}}, . \tag{2}$$

where $C_n(t)$ is the residue, whose form is fixed by the t-channel Regge trajectory (see[6]). The pole term in DAMA is a generalization of the Breit-Wigner formula, comprising a whole sequence of resonances lying on a complex trajectory $\alpha(s)$. Such a "reggeized" Breit-Wigner formula has little practical use in the case of linear trajectories, resulting in an infinite sequence of poles, but it becomes a powerful tool if complex trajectories with a limited real part and hence a restricted number of resonances are used.

A simple model of trajectories satisfying the threshold and asymptotic constraints is a sum of square roots[6] $\alpha(s) \sim \sum_i \gamma_i \sqrt{s_i - s}$. The number of thresholds included depends on the model; the simplest is with a single threshold

$$\alpha(s) = \alpha(0) + \alpha_1(\sqrt{s_0} - \sqrt{s_0 - s})\,. \tag{3}$$

Imposing an upper bound on the real part of this trajectory, $Re\,\alpha(s) < 0$, we get an amplitude that does not produce resonances, since the real part of the trajectory does not reach $n = 0$ where the first pole could appear. The imaginary part of such a trajectory instead rises indefinitely, contributing to the total cross section with a smooth background. This is the ansatz we suggest for the exotic trajectory[1] .

The super-broad-resonance approximation[7] , in a sense, is opposite to the narrow-resonance approximation, typical of the Veneziano model. However, contrary to the latter, valid only when the resonances widths vanish, the super-broad resonance approximation allows for a smooth transition to observable resonances or to the "Veneziano limit"[6] . Dual properties of this model were studied in[8] .

2. J/Ψ photoproduction

Photoproduction of vector mesons is well described in the framework of the vector meson dominance model[9] , according to which the photoproduction scattering amplitude A is proportional to the sum of the relevant hadronic amplitudes[10] (see also[11]):

$$D_H(\gamma P \to V P) = \sum_V \frac{e}{f_V} D_H(V P \to V P), \tag{4}$$

where $V = \rho, \omega, \phi, J/\Psi, \ldots$ Within this approximation, photoproduction is reduced to elastic hadron scattering ($J/\psi - p$ in our case), where the constants e and f_V are absorbed by the normalization factor, to be fitted to the data.

Among various vector mesons we choose J/ψ, because by the OZI rule[5] in $J/\psi - p$ scattering only the Pomeron trajectory can be exchanged in the t channel. To a lesser extent, this is true also for the $\phi - p$, however in the latter case ordinary meson exchange is present due to $\omega - \phi$ mixing. Heavier vector mesons are as good as $J/\Psi - p$, but relevant data are less abundant. So, we find $J/\Psi - p$ scattering to be an ideal testing field for diffraction, where it can be studied uncontaminated by secondary trajectories.[1,12]

In[1] we applied DAMA to meson-baryon scattering * with an exotic trajectory in the direct channel and the Pomeron trajectory in the exchange channel and calculate the differential and integrated elastic cross section for $J/\psi - p$ photoproduction.

*By having accepted vector dominance, we thus reduce J/Ψ photoproduction to $J/\psi - p$ scattering

The parameters of the model were fitted to the experimental data on J/Ψ photoproduction. With these parameters we calculated also the imaginary part of the forward amplitude proportional to the $J/\psi - p$ total cross section.

Let us briefly review the main steps in the calculations in[1] . Following[13] , we write the meson-baryon elastic scattering amplitude (with J/Ψ photoproduction in mind) as a combination

$$D(s,t,u) = (s-u)\,(D(s,t) - D(u,t))\,. \tag{5}$$

For the exotic Regge trajectory such as (3) the scattering amplitude, $D(s,t,u)$, is given by a convergent integral, eq. (5) with (1), and can be calculated for any s and t without analytical continuation, needed otherwise, as discussed in[6] .

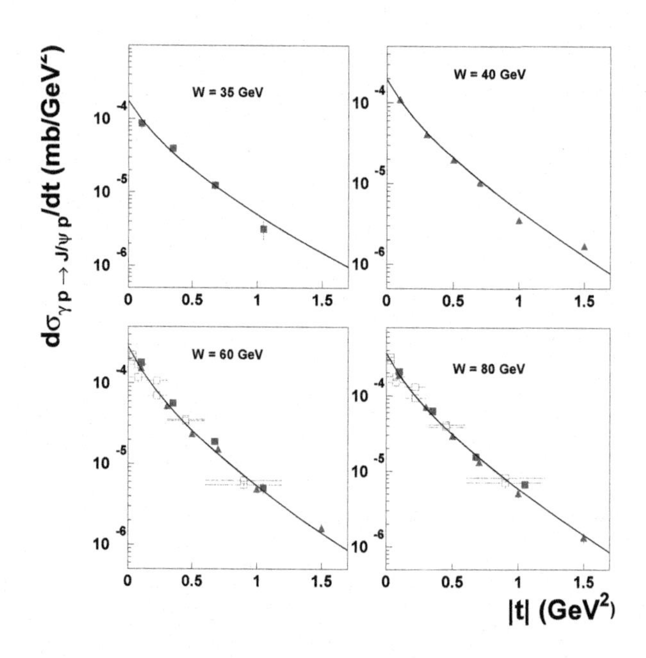

Fig. 1. J/Ψ differential cross sections as a function of t in the energy range W between 35 and 80 GeV.

We use a t-channel Pomeron trajectory in the form

$$\alpha(t) = \alpha^P(t) = \alpha^P(0) + \alpha_1^P(\sqrt{t_1} - \sqrt{t_1 - t}) + 2\alpha_2^P(t_2 - \sqrt{(t_2 - t)t_2}) \tag{6}$$

with a light (lowest) threshold $t_1 = 4m_\pi^2$ and a heavy one t_2, whose value, together with other parameters appearing in (1), are fitted to the data (see Table I).

The direct-channel exotic trajectory is given by eq. (3), where the relevant threshold value is $s_0 = (m_{J/\Psi} + m_P)^2$. Let us remind also that s, t and u are not independent variables but they are related by

$$s + t + u = 2m_{J/\Psi}^2 + 2m_P^2\,. \tag{7}$$

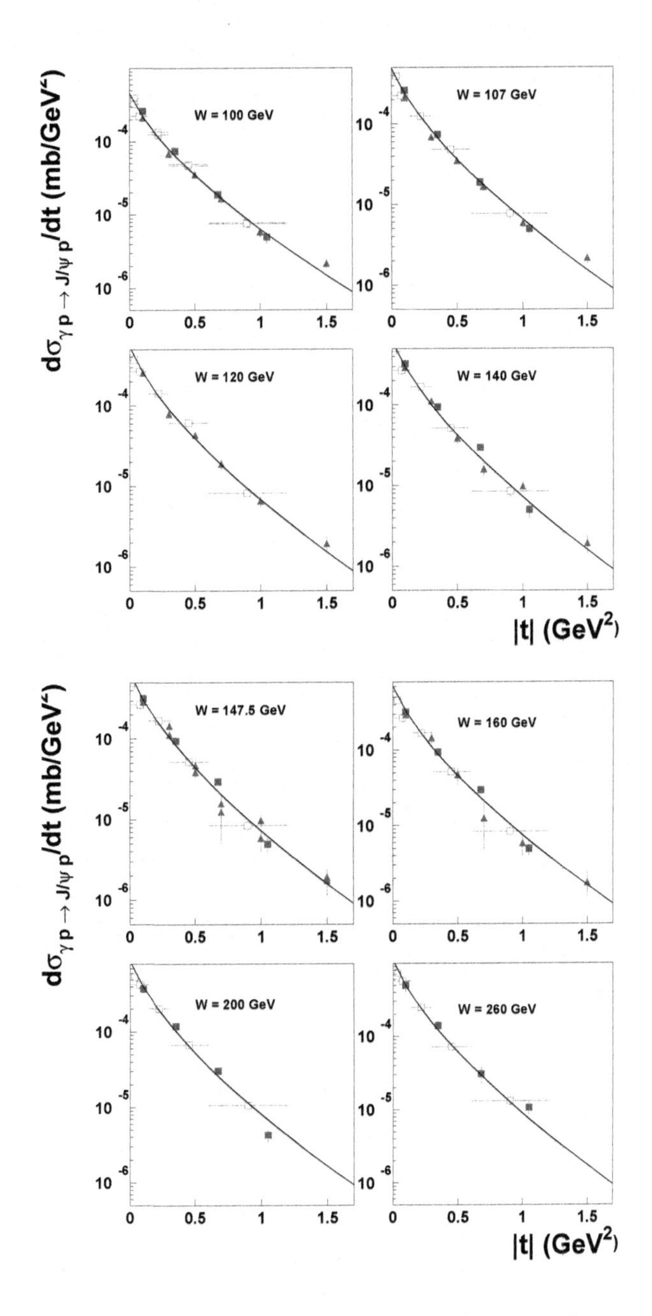

Fig. 2. J/Ψ differential cross sections as a function of t in the range of energy W from 100 to 260 GeV.

The differential cross section is then given by

$$\frac{d\sigma}{dt} = \frac{1}{16\pi\lambda(s, m_{J/\Psi}, m_P)} |D(s,t,u)|^2 , \tag{8}$$

Table 1. Fitted values of the adjustable parameters

$\alpha^E(0) = -1.83$	$\alpha_1^E(0) = 0.01$ (GeV^{-1})
$\alpha^P(0) = 1.2313$	$\alpha_1^P(0) = 0.13498$ (GeV^{-1})
$\alpha_2^P(0) = 0.04$ (GeV^{-2})	$t_2 =$ 36 (GeV^2)
$g =$ 13629	$c =$ 0.0025
	$\chi^2/d.o.f. =$ 0.83

where $\lambda(x, y, z) = x^2 + y^2 + z^2 - 2xy - 2yz - 2xz$.

The integrated elastic cross section is defined as

$$\sigma_{el}(s) = \int_{-t_{max}=s/2}^{t_{thr.}\approx 0} dt\, \frac{d\sigma}{dt}\,. \tag{9}$$

We have fitted the parameters of the model to the data[14,15] on J/Ψ photoproduction differential cross section in the energy range W between 35 to 260 GeV. The results of the fits together with the data are shown in Figs. 1 and 2. The values of the fitted parameters are quoted in Table 1.

Having fixed the parameters of the model we can now make a prediction for the integrated elastic cross section, which is shown in Fig. 3. The calculated curve is fully consistent with the data in the whole kinematical region. Enlarged is shown the energy region close to the threshold, where the background contribution dominates.

Thus, in[1] we have demonstrated the viability of the model in the whole kinematical region. Now we can ask at which energy the Regge behavior starts to dominate. The relevant energy can be extracted e.g. form the total cross section calculated from DAMA (with the parameters fitted to the data) divided by its asymptotic form $s^{\alpha^P(0)-1}$, as shown in Fig. 4. One can see that the ratio starts to deviate from 1 around 50 GeV, which means that below this energy down to the threshold non-asymptotic effects become important.

3. J/Ψ electroproduction

Now we would like to generalize our model for the off shell J/Ψ photoproduction (=electroproduction). Here there are two new important elements to be included into consideration.

A) We have to build the Q^2-dependent amplitude (Q^2 is a photon virtuality). This we can be done on the bases of the modified-DAMA formulas from[2] .

B) The longitudial component of the cross section starts to play an important role and it has to be taken into account. This can be done by using the parameterization proposed in the[16] .

Let us also note that the kinematic relation between the Mandelstam variables should be modified from eq. (7) to

$$s + t + u = 2M_N^2 + M_{J/\Psi}^2 - Q^2\,. \tag{10}$$

The (s, t, Q^2) term of a Modified Dual Amplitude with Mandelstam Analyticity

(M-DAMA)[2] is given by:

$$D(s,t,Q^2) = c\int_0^1 dz\left(\frac{z}{g}\right)^{-\alpha(s')-\beta(Q^{2\prime\prime})-1}\left(\frac{1-z}{g}\right)^{-\alpha_t(t'')-\beta(Q^{2\prime})}, \tag{11}$$

where $\beta(Q^2)$ is a monotonically decreasing dimensionless function of Q^2. In[2] the following form of $\beta(Q^2)$ has been proposed:

$$\beta(Q^2) = -\frac{\alpha_t(0)}{\ln g}\ln\left(\frac{Q^2+Q_0^2}{Q_0^2}\right). \tag{12}$$

Relating scattering amplitude for $t = 0$ to the nucleon structure function it was shown in[2] , that in both large and low x (Bjorken variable) limits as well as in the resonance region the parameterization (12) works well and the results of M-DAMA are in qualitative agreement with the experiment.

Thus, stepping from DAMA to M-DAMA brings in only one new parameters, namely the characteristic virtuality scale Q_0^2, see eq. (12).

The simplest form of the complete amplitude is given by a generalized eq. (5):

$$A(s,t,u,Q^2) = c(s-u)(D(s,t,Q^2) - D(u,t,Q^2)). \tag{13}$$

Now we can calculate $A(s,t,u,Q^2)$, using the same numerical method as in[1] .

In[2] the danger of having poles in Q^2 for the supercritical Pomeron trajectory is discussed. However, a more detailed study showed that we can use a supercritical Pomeron in M-DAMA without any problem[17] .

In this simple generalization procedure there is one strong assumption. The M-DAMA definition, eq. (11), as well as the form of the $\beta(Q^2)$, eq. (12), were developed and tested for the symmetric reactions, like $V(Q^2) + p \to V(Q^2) + N$, while in our case incoming J/Ψ is virtual while outgoing one is real. So, a more advanced model for virtual J/Ψ photoproduction amplitude, eq. (13), may be needed.

Now the total (integrated) elastic cross section is given by

$$\sigma_{el}(s,Q^2) = (1 + R(Q^2))\sigma_T(s,Q^2), \tag{14}$$

where $R = \sigma_L/\sigma_T$ ratio, and $\sigma_T(s,Q^2)$ is calculated according to the eq. (9) as before:

$$\sigma_T(s,Q^2) = \int_{-t_{max}=s/2}^{t_{thr.}\approx 0} dt\,\frac{d\sigma_T}{dt}(s,t,Q^2). \tag{15}$$

In[16] the following parameterization for the $R(Q^2)$ was proposed:

$$R(Q^2) = \left(\frac{Q_1^2+Q^2}{Q_1^2}\right)^n - 1, \tag{16}$$

where Q_1^2 and n are adjustable parameters.

Thus, in total we have 3 new adjustable parameters to fit the existing data on $\sigma_{el}(s,Q^2)$ J/Ψ photoproduction.

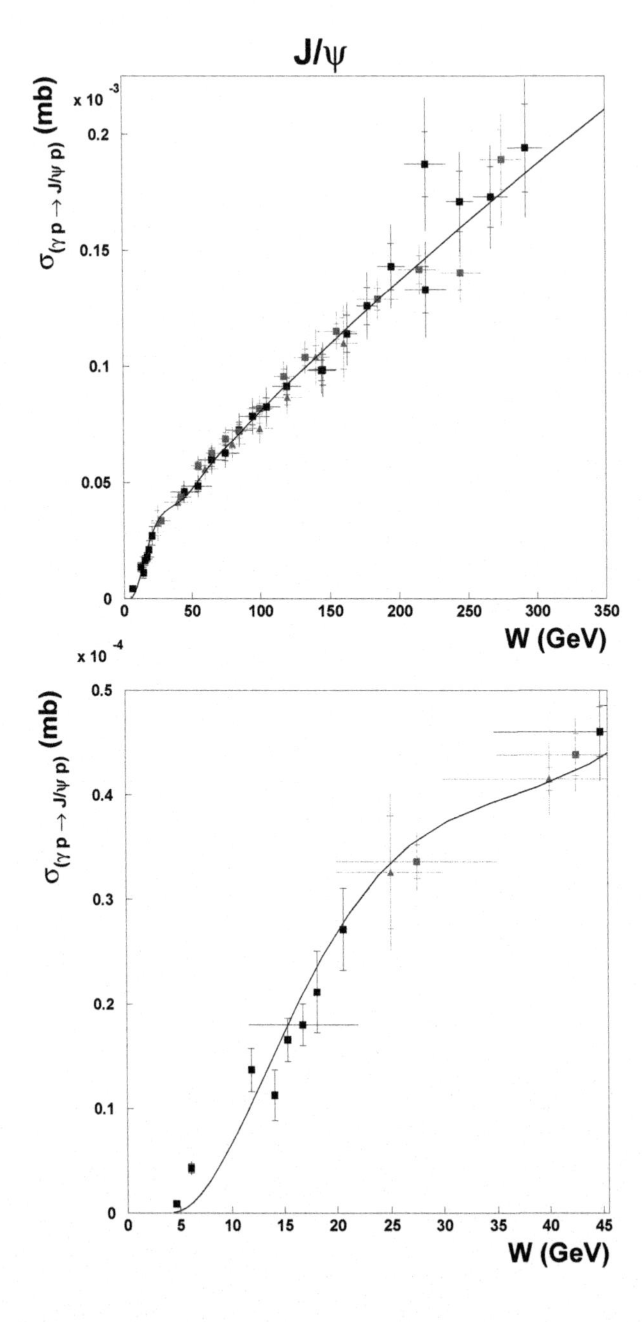

Fig. 3. J/Ψ elastic cross section for all energies (upper panel) and close to the threshold region (lower panel).

4. Conclusions

In this Letter we study J/Ψ photo- and electroproduction, which are the ideal testing field for diffraction, where it can be studied uncontaminated by secondary

References

1. R. Fiore, L. L. Jenkovszky, V. K. Magas, F. Paccanoni and A. Prokudin, *Phys. Rev. D* **75**, 116005 (2007).
2. V. K. Magas, *Phys. Atom. Nucl.* **68**, 104 (2005); hep-ph/0411335; hep-ph/0611119.
3. I. Ivanov, N. Nikolaev, and A. Savin, *Phys. Part. Nucl.* **37**, 1 (2006).
4. G. Veneziano, *Nuovo Cim. A* **57**, 190 (1968).
5. S. Okubo, *Phys. Lett.* **5**, 165 (1963); G. Zweig, *Preprints CERN* 401, 402, 412, CERN (1964); J. Iizuka, *Progr. Theor. Phys. Suppl.* **37-38**, 21 (1966).
6. A. I. Bugrij et al., *Fortschritte der Physik* **21**, 427 (1973).
7. A. I. Bugrij, L. L. Jenkovszky and N. A. Kobylinsky, Lett. *Nuovo Cim.* **5**, 389 (1972).
8. L. L. Jenkovszky, *Yadernaya Fizika*, (English transl.: *Sov. J. of Nucl. Phys.*) **21**, 213 (1974).
9. J. J. Sakurai, in *Proc. of the 1969 International Symposium on Electron and Photon Intaractions at High Enrergies, Liverpool 1969, ed. by D.W. Brades*, (Daresbury, 1969), p. 9; J. J. Sakurai and D. Schildknecht, *Phys. Lett. B* **40**, 121 (1972); J. Alwall, G. Ingelman, hep-ph/0310233.
10. P. D. B. Collins, *An Introduction to Regge Theory and High Energy Physics*, Cambridge University Press, 1977.
11. L. L. Jenkovszky, E. S. Martynov, F. Paccanoni, in *HADRONS-96, Novy Svet, 1996, edited by G.V. Bugrij, L.L. Jenkovszky and E.S. Martynov*, (Kiev, 1996), p. 170; A. Afanasiv, C. E. Carlson and Ch. Wahlqvist, *Phys. Rev. D* **61**, 034014 (2000).
12. L. L. Jenkovszky, S. Yu. Kononenko and V. K. Magas, hep-ph/0211158.
13. L. L. Jenkovszky, N. A. Kobylinsky, and A. B. Prognimak, *Ann. Phys.* **32**, 81 (1975).
14. A. Aktas *et al.* [H1 Collaboration], *Eur. Phys. J. C* **46**, 585 (2006).
15. S. Chekanov *et al.* [ZEUS Collaboration], *Eur. Phys. J. C* **24**, 345 (2002).
16. E. Martynov, E. Predazzi and A. Prokudin, *Phys. Rev. D* **67**, 074023 (2003); *Eur. Phys. J. C* **26**, 271 (2002).
17. V.K. Magas, in preparation.

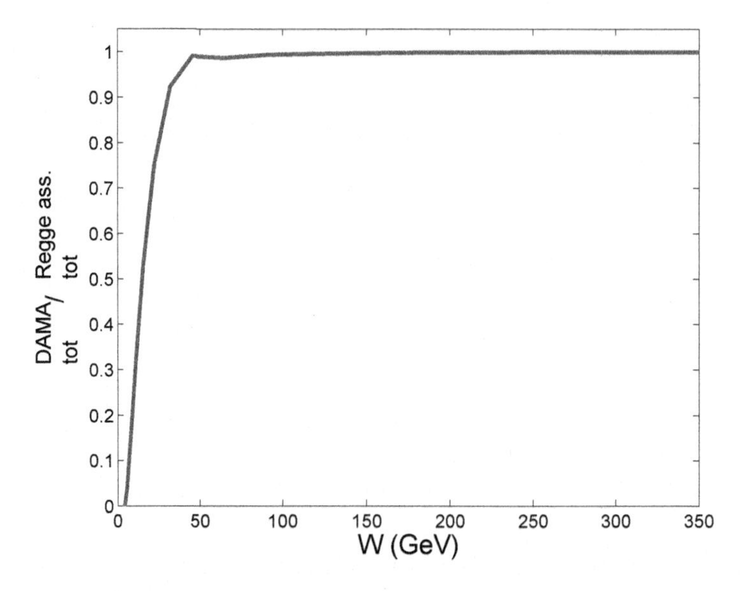

Fig. 4. J/Ψ total cross sections versus its Regge asymptotics $\sigma_{tot}^{Regge.ass.} \sim s^{\alpha^P(0)-1}$. Pure Regge asymptotic behaviour fails below $W = 50$ GeV.

trajectories[1,12] .

We have seen that the dual model[1] , based on DAMA and exotic trajectory (3), gives a very good description of the J/Ψ photoproduction differential and integrated elastic cross sections, measured in[14,15] . We have also shown that the Regge asymptotic behaviour starts to dominate at energies above $W_{Regge} = 50$ GeV, while below the scattering amplitude has to be calculated in a full dual model.

Also we present a way of generalizing this model for the J/Ψ electroproduction, based on the off mass shell extension of DAMA[2,17] .

More generally speaking, in this Letter and in[1] we introduced a new way of modeling background in different reactions, not only in J/Ψ photo- and electroproduction. The important new feature of our modeling is that background is "reggeized", i.e. it is parameterized in terms of exotic Regge trajectory. This trajectory can be of the type (3) or in the other form (not violating the DAMA constrains on the trajectory), but it has to satisfy the super-broad-resonance condition, $Re\,\alpha(s) < 0$[7] .

Acknowledgements

We thank Francesco Paccanoni for numerous valuable discussions on the subject of this paper. The work of L.J. was supported by the "Fundamental Properties of the Physical Systems at Extreme Conditions" program of the department of Astronomy and Physics, Ukrain's National Academy of Sciences.